Technology-Critical Elements

Metallic Pollutants in the Aquatic Environment

Rachel Ann Hauser-Davis
Oswaldo Cruz Foundation, Fiocruz
Brazil

CRC Press
Taylor & Francis Group
Boca Raton London New York

CRC Press is an imprint of the
Taylor & Francis Group, an **informa** business
A SCIENCE PUBLISHERS BOOK

Cover credit: Author created the images with the help of AI.

First edition published 2027
by CRC Press
2385 NW Executive Center Drive, Suite 320, Boca Raton FL 33431

and by CRC Press
4 Park Square, Milton Park, Abingdon, Oxon, OX14 4RN

CRC Press is an imprint of Taylor & Francis Group, LLC

Library of Congress Cataloging-in-Publication Data (applied for)

ISBN: 978-1-032-53493-0 (hbk)
ISBN: 978-1-032-53495-4 (pbk)
ISBN: 978-1-003-41232-8 (ebk)

DOI: 10.1201/9781003412328

Typeset in Palatino
by Prime Publishing Services

Preface

The role of metals in human advancement is profound and has been critical since the early discovery of copper around 9000 BC. Metals have significantly impacted various aspects of human activity, from transportation and agriculture to tools, weaponry, arts, construction, packaging, infrastructure, medical applications, and energy production. Historically, the "metals of antiquity"—gold, silver, copper, tin, lead, iron, and mercury—were fundamental to early civilizations, shaping their economic, cultural, and technological progress. Over time, the list of known metals has expanded considerably, now encompassing around 94 elements, depending on the employed classification (RSC, 2025). Modern society relies heavily on these metals, though many remain inadequately studied, particularly concerning their potential toxicity.

Metal contaminants can enter aquatic environments through various means, including industrial activities, agriculture, urban runoff, wastewater discharge, and accidental spills. These contaminants can have serious environmental consequences, such as the loss of biodiversity, disruptions to aquatic food webs, and overall ecosystem degradation. Additionally, they pose risks to human health through exposure to contaminated food items. This book series on metallic pollutants aims to provide a comprehensive exploration of lesser-studied metals, addressing their toxicity, effects on aquatic life, and potential health risks to humans.

This second book is on Technology-Critical elements, with particular emphasis on Rare Earth Elements (REEs), Rubidium (Rb), Platinum (Pt) and platinum group metals (PGMs), all of which are of critical importance to modern technological and industrial applications.

Rachel Ann Hauser-Davis

Contents

Chapter 1

Introduction

Technology-critical elements (TCEs) are chemical elements essential for modern and emerging technologies, playing an essential role in the production, performance, and advancement of a wide array of technological innovations across multiple industries, including electronics, renewable energy, healthcare, and transportation (Dang et al., 2021). These elements include rare earth elements (REEs), as well as other metals like cobalt, lithium, platinum and platinum-group metals, rubidium and indium, among others. They are also known as contaminants of emerging concern (CECs), comprising compounds whose environmental impact and potential risks to human health and wildlife have not been fully recognized or understood until recently (Feng et al., 2023).

The growing reliance on TCEs is deeply associated with their unique chemical, physical, and electronic properties, which make them indispensable across a wide range of industries. They are, for example, essential in sectors such as electronics, renewable energy, telecommunications, medical devices, aerospace, and defense, enabling the development of technologies like high-performance magnets, batteries, electric vehicles, smartphones, and advanced medical diagnostic tools (Dang et al., 2021; Abdou et al., 2023a). Consequently, the demand for TCEs has surged significantly in recent decades and is expected to continue as their importance in novel and advanced technologies increases.

A key factor behind the rising demand for TCEs is their role in advancing sustainable development goals, especially within the transition to low-carbon technologies. For example, TCEs are essential components in green technologies such as wind turbines, solar panels,

DOI: 10.1201/9781003412328-01

and electric vehicles (Baumgart et al., 2024; Trimmel et al., 2024), all of which are central to reducing carbon emissions and promoting energy efficiency. Moreover, these elements are also integral to energy storage systems, which are necessary for managing renewable energy generation and addressing the intermittency challenges of solar and wind power (Baumgart et al., 2024; Trimmel et al., 2024).

However, the criticality of TCEs is due not only from their technological importance but also from their scarcity and the geopolitical concentration of supply, which create significant vulnerabilities in the global supply chain (Filella, 2020). In this sense, a small number of countries control the majority of TCE production, which creates supply chain risks, especially in times of political instability, trade conflicts, or disruptions in market dynamics. This has increasingly raised concerns over resource security for many nations, as they increasingly depend on these materials for their technological and industrial needs. As a result, several countries are now exploring alternative sources, increasing recycling efforts, and establishing strategic stockpiles to mitigate potential shortages and ensure a reliable supply (Nuss and Blengini, 2018).

Alongside these supply chain vulnerabilities, the extraction and processing of TCEs present serious environmental challenges (Filella and Rodríguez-Murillo, 2017; Bu-Olayan and Thomas, 2020). Mining and refining TCEs often require the use of toxic chemicals, high energy consumption, and the generation of large amounts of waste, which can cause significant harm to ecosystems and communities. Issues such as water pollution, soil contamination, and the release of radioactive byproducts are among the environmental risks associated with TCE extraction. As the demand for TCEs continues to grow, addressing the environmental impact of their extraction, processing and use becomes more urgent. This includes improving mining techniques, advancing recycling technologies, and developing cleaner, more sustainable practices for extracting and refining these elements.

THE COSTS OF MINING AND USING TECHNOLOGY-CRITICAL ELEMENTS

The unsustainable mining, processing and disposal of TCEs can have potentially significant harmful environmental and human health effects (Ali and Katima, 2020). In a recent advisory report, the Scientific and Technical Advisory Panel (STAP) of the Global Environment Facility (GEF) has discussed these negative effects (Ali and Katima, 2020), indicated below.

Concerning environmental costs, the extraction of TCEs often requires extensive land clearance and deforestation, directly leading to habitat loss, land degradation, and biodiversity decline. For instance, it is estimated that up to 300 m^2 of vegetation and topsoil are removed for every tonne of rare earth oxide extracted. Forested regions, such as the Amazon and the Democratic Republic of Congo, where biodiversity is exceptionally high, are particularly vulnerable to these impacts. Mining activities near or within protected areas exacerbate the risks to ecosystems, threatening species survival and ecosystem integrity. Mining operations in desert ecosystems, while traditionally considered less harmful to biodiversity, also pose unique challenges due to water scarcity. The substantial water requirements for mineral processing can strain already limited water resources in these regions, creating conflicts over water use and threatening local communities and ecosystems.

Furthermore, mining operations for TCEs can lead to the formation of acid mine drainage, a phenomenon where sulfide minerals are exposed to air and water, forming sulfuric acid. This acidification disrupts marine and freshwater ecosystems, killing aquatic organisms and reducing biodiversity. Additionally, acid mine drainage can contaminate soil and water resources, with long-term impacts on surrounding ecosystems. Mining operations also release significant amounts of dust containing TCEs and other toxic substances into the air. Activities such as cutting, drilling, blasting, and open burning of electronic waste (e-waste) exacerbate this issue, exposing nearby wildlife, vegetation, and human communities to harmful pollutants.

Following extraction, processing TCEs generates vast quantities of waste. For every ton of rare earth oxide extracted, up to 1,000 tons of contaminated wastewater and 2,000 tons of tailings are produced. These by-products often contain toxic substances and are frequently disposed of in adjacent valleys or streams, leading to contamination of surface and groundwater. The flow of these pollutants into international waters can create transboundary environmental challenges. The extraction and processing of TCEs often require significant energy inputs, typically derived from fossil fuels. This results in substantial greenhouse gas emissions. Additionally, the removal of forests for mining operations eliminates carbon sinks, further exacerbating climate change.

Furthermore, as a consequence of their rapidly growing use over the past few decades, TCEs are now being increasingly detected across a wide range of environmental matrices worldwide. Aquatic environments, in particular, have emerged as major sinks for these contaminants, largely due to the inadequate management and disposal of industrial effluents, urban runoff, and e-waste, a significant source of TCEs given their high content of components such as circuit boards, magnets, batteries, and display technologies. This widespread release of TCEs into natural systems raises serious concerns about their long-term persistence, bioaccumulation potential, and possible impacts on ecosystem and human health. Currently, only about 20% of e-waste is recycled. As global demand for electronic goods rises, the volume of e-waste is expected to grow, amplifying the environmental release of TCEs.

With regard to human health effects, workers in TCE mining and processing operations are often exposed to harmful chemicals and particulate matter. Prolonged exposure to these has been linked to respiratory illnesses, including pneumoconiosis, and diseases such as lung cancer. Chemicals used in TCE ore extraction and refining are also often inadequately regulated, resulting in water pollution and adverse health effects for nearby communities. Furthermore, many TCEs, such as uranium and cadmium, contain radioactive or toxic elements. Improper disposal and waste management can lead to air, water, and soil contamination, exposing humans to hazardous materials and increasing the risks of cancer and other chronic diseases.

Finally, many TCEs released into the environment can enter food chains through soil, water, and vegetation. Biomagnification and potentially biomagnification of these elements can lead to significant health risks for humans, including neurological damage, kidney dysfunction, and developmental issues.

However, despite the widespread use of TCEs, significant knowledge gaps remain regarding their anthropogenic levels, environmental behavior, and toxicological effects, significantly hindering full understanding of their long-term impacts (Dang et al., 2021; Kouhail et al., 2022). Therefore, further research is required to elucidate the biogeochemical cycling of TCEs, as well as their environmental fate and cumulative exposure effects. Additionally, the development of sustainable mining practices, recycling technologies, and effective waste management strategies, as well as the development of alternative materials, is critical to mitigating these impacts.

TECHNOLOGY-CRITICAL ELEMENT BEHAVIOR IN AQUATIC ENVIRONMENTS

The speciation and bioavailability of TCEs in aquatic environments are strongly influenced by a range of physicochemical water properties, such as salinity, pH, redox potential, temperature, and the presence of organic and inorganic ligands.

Salinity is particularly important in marine systems, where high concentrations of chloride and sulfate ions promote the formation of stable metal-ligand complexes. Several TCEs often form anionic chlorocomplexes under saline conditions, which are typically less bioavailable because their negative charge limits passive diffusion across biological membranes (Khan et al., 2017; Gantayat and Elumalai 2024). Similarly, some TCEs can form stable complexes with carbonate and sulfate, decreasing their free ionic forms and thus reducing their potential uptake by organisms (Khan et al., 2017; Batley and Campbell, 2022).

pH also plays a critical role by influencing both metal speciation and the availability of key ligands. In the relatively alkaline pH of seawater (~8.1), several TCEs tend to form carbonate complexes, lowering their bioavailability (Smrzka et al., 2019; Gantayat and Elumalai 2024). However, in more acidic conditions, such as those resulting from localized pollution or ongoing ocean acidification, these carbonate complexes may break down, increasing the proportion of free TCE ions, which are more readily taken up by marine biota (Khan et al., 2017; Romero-Freire et al., 2019a).

Redox conditions further influence the oxidation state and mobility of metals and metalloids, including TCEs. In oxygenated surface waters, some TCEs are more likely to be found in soluble, oxidized forms, whereas in anoxic environments, such as sediments or hypoxic zones, reduced species may dominate or even precipitate, reducing their solubility and bioavailability (Miao et al., 2006; Khan et al., 2017). The redox state also determines whether metals remain in the dissolved phase or bind to particles and organic matter (Borch et al., 2010; Madadi et al., 2023).

The presence of dissolved organic matter also directly affects TCE forms and bioavailability in aquatic environments, as organic ligands such as humic substances, peptides, and thiols can bind with metals to form organometallic complexes (Zhao et al., 2016; Pontoni et al., 2022). These complexes may either enhance or reduce metal bioavailability depending on their size, solubility, and the specific transport mechanisms of marine organisms.

Temperature can also indirectly affect speciation and directly influence TCE uptake by altering the kinetics of complexation reactions and the metabolic rates of organisms. Higher temperatures often increase the rate at which metals dissolve or react with ligands, as well as the efficiency of active transport systems in marine animals (Hoffmann et al., 2012). Additionally, temperature shifts may influence the equilibrium between different chemical species, potentially altering their environmental mobility and biological effects (Kazmi et al., 2022). This is particularly concerning in the current climate-change scenario.

Finally, the presence of major ions in seawater, such as sodium, potassium, calcium, and magnesium, can compete with trace elements for uptake pathways in marine organisms (Wright, 1995). These ions play critical and highly conserved roles in maintaining cellular and physiological homeostasis in virtually all marine organisms. As many TCEs share similar ionic radii or charge densities with these essential elements, they can mimic their behavior, allowing them to enter cells via the same transport systems. Such competitive interactions can disrupt ion homeostasis, interfering with essential cellular functions, nutrient homeostasis, and ion transport mechanisms, significantly and negatively affecting organismal health and ecosystem dynamics, *i.e.*, leading to decreased growth rates, reproductive failure, and even increased mortality, especially in early developmental stages that are more vulnerable to ionic imbalances.

The next sections refer to on the properties, uses, and potential risks of Rare Earth Elements (REEs), rubidium (Rb) and platinum and platinum-group metals (PGMs) as some of the most representative TCEs currently in use. Discussions on their roles in modern technology and the growing concerns over their sustainability and human and environmental health impacts are conducted, focusing on aquatic ecosystems.

Chapter 2

Rare Earth Elements

Rare Earth Elements (REEs) comprise a group of 17 chemical elements, including the 15 lanthanides—members of the 4f block with atomic numbers ranging from 57 to 71—alongside scandium (Sc) and yttrium (Y). The lanthanide group consists of lanthanum (La), cerium (Ce), praseodymium (Pr), neodymium (Nd), promethium (Pm), samarium (Sm), europium (Eu), gadolinium (Gd), terbium (Tb), dysprosium (Dy), holmium (Ho), erbium (Er), thulium (Tm), ytterbium (Yb), and lutetium (Lu). Scandium and yttrium are included among the REEs due to their occurrence in the same deposits as lanthanides and their similar chemical properties (Massari and Ruberti, 2013).

The term "Rare Earths" is, however, somewhat misleading. It originated due to the perceived rarity of the minerals from which these elements were first isolated in their oxide forms (BGS, 2011). The designation "earths" refers to their oxide form, which has a soil-like appearance (Miranda Jr, 2000). Although these elements were once considered scarce, largely due to the absence of advanced purification technologies in the past (Miranda Jr., 2000), they are now recognized as relatively abundant in the Earth's crust, with an average concentration of 9.2 parts per million (ppm) (Rudnick et al., 2005). Despite this abundance, these elements are seldom found in highly concentrated deposits, which significantly complicates their extraction and purification processes. Their diffuse distribution within geological formations increases the technical and economic challenges associated with isolating these elements in usable forms, requiring sophisticated mining, processing, and refining technologies. This combination of relative abundance and distribution complexity highlights their critical role in modern industry and the ongoing need for innovation in extraction methodologies (Rudnick et al., 2005).

DOI: 10.1201/9781003412328-02

THE DISCOVERY OF RARE EARTH ELEMENTS

The history of REEs traces its roots back to the 18th century, with early explorations beginning in 1751 when Swedish mineralogist and chemist Axel Frederik Cronstedt first identified a new mineral that he named cerite (Evans, 2012). Initially, Cronstedt believed the mineral to be a silicate of iron and calcium, but it later became evident that cerite contained a previously unknown suite of elements. This early discovery marked the beginning of the journey toward identifying and understanding REEs, although it would take many more years before the chemistry of these elements would be fully developed.

The real cornerstone of REE chemistry is often attributed to Carl Axel Arrhenius, a Swedish mineralogist, chemist, and military officer (Adunka and Orna, 2018). In 1787, Arrhenius discovered gadolinite, a mineral which was initially referred to as ytterbite. Gadolinite was composed of a complex mix of silicates, including those of cerium, lanthanum, neodymium, yttrium, beryllium, and iron (Rocio et al., 2012; Sastri et al., 2003). The mineral's importance was recognized as it contained a variety of unfamiliar elements that would later be categorized as REEs.

The next significant breakthrough occurred in 1794 when Johan Gadolin, a Finnish chemist, is credited with isolating yttrium from gadolinite (Cesbron, 1989). Gadolin's work, particularly his isolation of yttrium, established him as one of the key figures in the development of REE chemistry. Yttrium was initially identified as a new chemical element, making it one of the first REEs to be isolated. Gadolin's work laid the foundation for further exploration into the properties and potential uses of these enigmatic elements, earning him recognition as the father of REE chemistry (Sousa Filho and Serra, 2014).

As research on REEs progressed throughout the 19th and early 20th centuries, several other key elements were isolated and named. However, it was not until the early 1900s that the last naturally occurring REE, lutetium, was discovered in 1907 (Pöttgen et al., 2020). The discovery of this element was marked by a scientific dispute among three prominent scientists: Georges Urbain, a French chemist; Carl Auer von Welsbach, an Austrian mineralogist; and Charles James, an American chemist. Urbain argued that ytterbium, another rare earth element, could be split into two distinct elements, one of which he called lutetium. Welsbach and James independently made similar claims, leading to a heated controversy regarding priority. The dispute was eventually settled in 1909 when the International Commission on Atomic Weights recognized Georges Urbain as the primary discoverer of lutetium. Urbain had been the first to publish his findings independently, leading to the formal attribution of the discovery to him (RSC, 2024a). Lutetium's discovery marked the final addition to the naturally occurring REEs, bringing the number of known REEs to 17.

RARE EARTH ELEMENT PROPERTIES AND CHARACTERISTICS

Rare Earth Elements are typically categorized into two main subgroups based on their atomic weight and chemical behavior as light REEs (LREEs) and heavy REEs (HREEs) (Voncken, 2016). The LREEs consist of the first half of the lanthanide series, spanning from La to Eu, while HREEs comprise the latter part of the lanthanide series, from Gd to Lu and also include Y. Although Y is lighter than the lanthanides, it is classified alongside heavy REEs due to its chemical and physical similarity to the heavier elements found in natural deposits (Lapido-Loureiro, 2013). Heavy REEs are generally less abundant in the environment (Balaram, 2019), and thus rarer and more difficult to extract, making them more expensive and strategically important (Talan and Huang, 2022).

These elements comprise two main groups, lanthanides and actinides, based on their electron configurations and position in the periodic table (Zepf, 2013). Both are characterized by the progressive filling of f-orbitals in their electron configurations, which influences their unique magnetic, optical, and chemical properties (Voncken, 2016).

The names lanthanides and actinides are derived from the first element in each respective series—lanthanum (La) and actinium (Ac)—and reflect their chemical and structural similarities. The term "lanthanide" comes from the Greek word *lanthanein* (λανθάνειν), which means "to lie hidden" or "to escape notice.", probably coined by Victor Goldschmidt, a pioneering German mineralogist and geochemist who made significant contributions to the field of geochemistry and is often regarded as one of the founders of modern geochemistry, in 1952 (Hakala, 1952). This name was chosen because lanthanum, the first element in the series, discovered in January 1839 by Carl Gustav Mosander at the Karolinska Institute, Stockholm, who extracted it from cerium (Ce) (RSC, 2024b), and its related elements were difficult to isolate and identify due to their similar chemical properties, and REE typically tend to "hide" behind other minerals. Similarly, the name "actinide" is derived from actinium (Ac), discovered in 1899 by André Debierne at Paris, who extracted it from the uranium ore pitchblende (uranium oxide, U_3O_8) (RSC, 2024c). The suffix "-ide" was added by analogy with lanthanides, signifying that these elements share chemical properties with actinium. The name "actinium" itself comes from the Greek word *aktinos* (ἀκτίνος), meaning "ray" or "beam," due to its strong radioactivity and ability to emit energetic radiation (RSC, 2024c).

Lanthanides consist of 15 elements that are categorized as inner transition metals due to the filling of the 4f sublevel in their electron configurations. In lanthanides, the highest-energy electrons are located in the 4f orbital, which is an inner shell, as opposed to the outermost shell, or valence shell, seen in many other elements. This unique characteristic leads to important similarities among these elements, such as their ability to form a stable +3 oxidation state and their closely related ionic radii. Because of these features, a single mineral can often contain multiple REEs. Similarly, actinides, which fill the 5f sublevel, share the same inner transition metal classification. The electron configurations of both lanthanides and actinides contribute to their similar chemical and physical behaviors, distinguishing them from elements with higher-energy electrons occupying their outermost shell (Johansson, 2000). This unique electron arrangement positions lanthanides and actinides apart from other elements in the periodic table, granting them distinct properties. Table 2.1 depicts some REE properties, such as atomic number, atomic weight, ground state configuration and categorization as light or heavy.

Table 2.1 Rare earth element atomic numbers, atomic weights, ground state configurations and categorization as light or heavy (adapted from Aide and Aide, 2012; Harmer and Nex, 2016)

Element	Atomic Number	Atomic Weight	Ground State Configuration	Categorization
La	57	138.90	[Xe] $5d^1$ $6s^2$	Light REE
Ce	58	140.12	[Xe] $4f^1$ $5d^1$ $6s^2$	
Pr	59	140.90	[Xe] $4f^3$ $6s^2$	
Nd	60	144.24	[Xe] $4f^4$ $6s^2$	
Pm	61	145.00	[Xe] $4f^5$ $6s^2$	
Sm	62	150.36	[Xe] $4f^6$ $6s^2$	
Eu	63	151.96	[Xe] $4f^7$ $6s^2$	
Gd	64	157.25	[Xe] $4f^7$ $5d^1$ $6s^2$	Heavy REE
Tb	65	158.92	[Xe] $4f^9$ $6s^2$	
Dy	66	162.50	[Xe] $4f^{10}$ $6s^2$	
Ho	67	164.93	[Xe] $4f^{11}$ $6s^2$	
Er	68	167.26	[Xe] $4f^{12}$ $6s^2$	
Tm	69	168.93	[Xe] $4f^{13}$ $6s^2$	
Yb	70	173.04	[Xe] $4f^{14}$ $6s^2$	
Lu	71	174.96	[Xe] $4f^{13}$ $5d^1$ $6s^2$	

Because of their electron configuration, lanthanides share certain similarities that set them apart from other elements. One of the most prominent features is their +3 oxidation state, which is thermodynamically stable across the entire group. This common oxidation state results from the removal of three electrons, two from the 4f sublevel and one from the 5s sublevel. As a result, lanthanides typically form trivalent cations (Ln^{3+}) in compounds, making them behave in very similar ways chemically (Johansson, 2000).

Lanthanides also exhibit very close ionic radii, which is another defining characteristic of this group. The ionic radii of these elements are nearly identical due to the filling of the 4f orbitals, where the additional electrons do not significantly alter the overall size of the ion (Johansson, 2000). This similarity in size means that lanthanides tend to substitute for each other in minerals, as their ionic radii are close enough to make them interchangeable in many cases. In fact, a single mineral deposit can often contain a mixture of several different but chemically similar lanthanides, making them difficult to separate from one another without applying advanced techniques (Giese et al., 2016; Castor and Hendrick, 2006; Abrão, 1994), dictating the type of mining process employed (Denton et al., 2019).

This unique electron configuration and the resulting similarities in physical and chemical properties play a significant role in the applications of lanthanides in various high-tech industries. Their use in magnets, phosphors, catalysts, and battery technologies stems, in part, from the ability of these elements to function similarly in a range of chemical processes and technological applications. However, the shared properties also present challenges in terms of their extraction and purification, as separating them from one another often requires complex and costly processes.

After refining, REEs exhibit several interesting physical and chemical properties, as they are malleable, ductile, and reactive, with a silvery-gray appearance and high melting points (Halka and Nordstrom, 2019). These melting points increase progressively across the series, ranging from 798°C for Ce to 1663°C for Lu. Additionally, their unique electron configurations impart magnetic and optical properties to these elements (Gupta and Krishnamurthy, 1992). These specific properties, particularly their magnetic and spectroscopic characteristics, have made REEs indispensable in various fields, including chemistry, electronics, defense, and technology. Their applications range from the development of advanced materials and renewable energy technologies to critical uses in military equipment and high-performance devices (Sousa Filho and Serra, 2014).

NATURAL OCCURRENCE AND GLOBAL DISTRIBUTION OF RARE EARTH ELEMENTS

Rare Earth Elements are naturally distributed across a range of countries worldwide, with significant deposits found in regions such as Australia, Brazil, Canada, China, the United States, India, Malaysia, and Russia (Denton et al., 2019). China is currently the largest global REE producer, holding most of the world's reserves and controlling a significant portion of the market (Dushyantha et al., 2020). In fact, China's dominance in the REE industry has raised geopolitical concerns due to its strategic control over the supply of these critical elements. Other countries, such as Australia and the United States, also have significant REE deposits, with Australia being the second-largest producer and the United States possessing some of the most promising, albeit underdeveloped, sources of these elements (Fernandez, 2017). Brazil is emerging as a potential player in the global REE market, with its own mineral REE reserves, primarily concentrated in the states of Minas Gerais and Bahia. In addition, Russia and India are home to notable REE deposits, with Russia focusing on increasing its domestic production, and India exploring its own untapped potential. Other countries like Malaysia and various African nations also hold smaller but potentially important sources of REEs, contributing to the growing diversification of the supply chain (Fernandez, 2017). The global distribution of REEs across these countries highlights the strategic importance of these elements for a wide range of technological applications and highlights the geopolitical and economic dynamics that influence their extraction, trade, and consumption.

These elements are primarily hosted in rock-forming minerals, including igneous, sedimentary, and metamorphic rocks, where they commonly substitute for major ions in the mineral lattice. The concentration and distribution of REEs are influenced not only by the type of rock formation but also by the geological processes that govern their formation, mobilization, and enrichment.

Deposits of REEs are broadly categorized into two main types: primary deposits, which are directly associated with igneous processes, such as magmatic differentiation where, during the cooling and crystallization of magma, REEs are often concentrated in late-stage residual melts, leading to the formation of REE-rich pegmatites and alkaline complexes (Chakhmouradian and Zaitsev, 2012; Anenburg et al., 2021) and hydrothermal activity, where hot, mineral-rich fluids can transport REEs and precipitate them in hydrothermal veins and deposits (Olivarez and Owen, 1989) and secondary deposits, which result from sedimentary processes, including the concentration of REEs through weathering, erosion, and deposition in sedimentary basins.

Secondary deposits are often enriched in REEs due to the leaching and redistribution of elements during prolonged weathering events, especially in tropical climates (BGS, 2011). Further classification of REE deposits is based on genetic associations, mineralogical characteristics, and their mode of occurrence. For instance, some deposits are linked to carbonatites and alkaline rocks, which are typically enriched in light REEs, while others, such as ion-adsorption clays found in weathered crusts, are rich in heavy REEs (BGS, 2011).

Heavy REEs (HREEs) are commonly found in minerals such as gadolinite, xenotime, samarskite, euxenite, and fergusonite. In contrast, light REEs (LREEs) are predominantly located in bastnäsite, monazite, loparite, and allanite (Rocio et al., 2012). Over 250 minerals are known to contain REEs, with approximately 55 being more abundant in the Earth's crust. Among these, monazite, xenotime, and bastnäsite are the most important for industrial purposes, sources of light and heavy REEs, respectively, although REEs are also found in apatite and other minerals (Sousa Filho and Serra, 2014; Jordens et al., 2013). They are also sometimes retained in lateritic clays and found in marine and alluvial sandy deposits, known as placers, formed through the physical weathering, erosion, transport, and deposition of REE-bearing minerals from their original rock sources into riverbeds, beaches, shallow marine environments, and alluvial plains (Babu et al., 2024).

RARE EARTH ELEMENT EXTRACTION AND PROCESSING

The extraction of REEs presents significant financial and technical challenges, which contribute to the complexity of developing economically viable operations for these valuable minerals. One of the primary obstacles is the high cost associated with the extraction process, which involves not only the initial discovery and identification of REE-rich deposits but also the subsequent extraction, processing, and refining stages. Unlike more traditional mining operations, REEs are rarely found in large, concentrated deposits, and are often embedded within other minerals (Batapola et al., 2020). This means that significant volumes of ore must be mined to obtain even small amounts of REEs, resulting in high operational costs and long processing times. The low concentration of REEs in many of the world's deposits further exacerbates these difficulties, making it financially unfeasible to extract and process them unless there is a consistent demand that justifies the investment (CGEE, 2013).

After the extraction phase, the REE-bearing ores are typically processed, often directly at the mining site to reduce transportation costs (Weng et al., 2015). The processing involves the use of both physical and chemical methods to separate REEs from other minerals and impurities in the ore (For a detailed review, please see Chen et al., 2022). Briefly, initially, physical techniques such as crushing, grinding, and flotation are employed to concentrate the REE-bearing minerals. Once the REEs are isolated from the bulk of the ore, more complex chemical processes are used to purify and separate individual elements. These methods can involve leaching with acid or base solutions, solvent extraction, ion exchange, or selective precipitation (BGS, 2011). These chemical processes are highly specialized, require significant energy inputs, and generate waste materials, some of which may be hazardous. As a result, the environmental and economic costs of processing REE ores can be considerable, further adding to the challenge of making REE extraction financially viable. Despite these obstacles, the increasing demand for REEs in high-tech industries, renewable energy technologies, and defense sectors is driving extensive efforts to improve the efficiency, sustainability, and security of REE mining, extraction, and processing.

THE SPECIFIC CASE OF RARE EARTH ELEMENT OCEAN EXPLORATION

With the global race to secure raw materials for the energy transition and the potential decline in the international supply of REEs, several countries have turned their attention to the oceans as a potential new frontier for REE exploration. This initiative seeks to determine whether REEs exist in marine environments and whether their extraction from the seabed is viable. However, all extraction and mining activities must first be authorized by the International Seabed Authority (ISA), a specialized United Nations body established in Jamaica. Created in 1994 under the framework of the United Nations Convention on the Law of the Sea (UNCLOS), the ISA plays a central role in regulating and organizing activities related to the use of oceanic mineral resources, ensuring that such operations are conducted in a manner that benefits all of humanity while safeguarding the health of the marine environment. The ISA also promotes scientific research, mining, and environmental protection in these areas (ISA, 2024).

The ISA's primary mandate, as outlined in UNCLOS, involves the regulation and oversight of all activities related to the exploration and exploitation of mineral resources in the international seabed area, which covers the seabed and subsoil areas beyond the jurisdiction of any single nation, extending to the ocean floor that lies outside of national economic zones, generally beyond 200 nautical miles from the coastlines of any country and lying outside the exclusive economic zones of individual nations.

The ISA's role is both regulatory and custodial, ensuring that the use of seabed resources is carried out in accordance with international legal frameworks and environmental standards. It is entrusted with the responsibility of managing these resources as the "common heritage of mankind," a concept enshrined in UNCLOS that emphasizes equitable access and shared benefits. A key aspect of the ISA's mandate is to ensure that the development of deep-sea mineral resources does not disproportionately favor industrialized nations but rather contributes meaningfully also to the economic advancement of developing countries, and contributing to global prosperity (Lodge et al., 2017). To this end, the ISA facilitates technology transfer, promotes scientific research and environmental baseline studies, and requires contractors to submit detailed plans of work and environmental impact assessments before proceeding with mineral exploration or exploitation.

Furthermore, the ISA is tasked with establishing and enforcing rules, regulations, and procedures designed to prevent damage to the marine environment and to safeguard biodiversity in the deep sea. This includes the creation of environmental management plans and the designation of areas of particular environmental interest (APEIs) to be set aside from mining activities.

To date, the ISA has focused its marine spatial planning primarily on individual exploration contract blocks. However, effective environmental management requires broader measures, including robust baseline studies, impact assessments, post-impact monitoring, and mechanisms to address harmful effects (Tunnicliffe et al., 2020). To support clear environmental objectives and promote the sustainable use of ocean resources, some authors recommend that planning be expanded beyond individual sites through the development and implementation of a Strategic Environmental Management Plan (SEMP) at a regional scale (Tunnicliffe et al., 2020). These would include several key objectives to effectively meet environmental goals in the context of deep-sea mining, as follows (Tunnicliffe et al., 2020):

- Ecosystems must be protected from contamination by pollutants generated during any phase of the mining process.
- It is essential to maintain the ability of populations to replace themselves, which includes preserving habitat quality and ensuring connectivity among populations.
- Ecosystem functions, such as natural productivity, nutrient cycling, and trophic relationships, should be safeguarded to prevent long-term degradation.

- Additionally, efforts must be made to avoid significant losses in genetic diversity, species richness, community structure, and habitat types, all of which contribute to ecological stability.
- The protection of ecosystem services, including those that remain poorly understood like carbon sequestration, is also crucial.
- Finally, maintaining the resilience of marine ecosystems is vital to prevent regime shifts and support recovery from cumulative impacts, including those from mining activities that may disrupt source populations, migratory pathways, life-history traits, and species distributions.

Deep-sea REE deposits are found in the form of different geological formations. These include ferromanganese nodules, found primarily in abyssal plains, which contain significant concentrations of REEs, along with cobalt, nickel, and manganese and marine sediments, particularly in the Pacific Ocean, which contain high concentrations of REEs due to the gradual accumulation of weathered materials from landmasses, due to their strong sorption capacity, which allows them to accumulate significant quantities of these valuable elements over time (Toyoda et al., 1990). Sediment samples from hydrothermal fields along the Mid-Atlantic Ridge and Pacific Ocean reveal significant variation in REE concentrations. Fields like Logatchev and Saldanha show lower levels, likely due to factors such as lower fluid temperatures and short fluid interactions with seawater. In contrast, regions like the NW Pacific contain highly enriched REE deposits, with some sediment cores surpassing 5,000 mg kg^{-1} (Milinovic et al., 2021).

Hydrothermal vent deposits are also noteworthy REE deposits. These formations emit metal-laden fluids that precipitate valuable minerals, including REEs, onto surrounding substrates, with LREE, particularly Eu and Ce, enriched due to their higher stability in chloride and fluoride complexes (Milinovic et al., 2021). Finally, ferromanganese crusts, found on underwater seamounts and ridges, are usually enriched with REEs, platinum-group metals, and cobalt.

Other REE-rich resources in the ocean include polymetallic nodules, cobalt-rich Fe-Mn crusts, and seafloor massive sulfides. Polymetallic nodules are mineral deposits found on the ocean floor, particularly in the abyssal plains at depths ranging from 4,000 to 6,000 meters (Elderfield et al., 1981; Milinovic et al., 2021). These nodules are composed of a mixture of metals, including manganese, nickel, cobalt, and REEs, along with smaller amounts of copper, zinc, and iron. The Clarion-Clipperton Zone (CCZ) in the central Pacific Ocean is the most well-known and explored region for polymetallic nodule deposits, with estimates suggesting there could be billions of tons of nodules across this vast area (Elderfield et al., 1981; Milinovic et al., 2021).

Ferromanganese crusts are often enriched in REEs, particularly in cobalt-rich deposits, which form over millions of years on the flanks of underwater mountains, seamounts, and ridges. These crusts are typically found at depths of 200 to 4,000 meters on seamounts and ridges in the Pacific Ocean, Atlantic Ocean, and Indian Ocean (Milinovic et al., 2021; Vysetti, 2023).

Seafloor massive sulfides typically contain significant amounts of REEs in addition to other metals, with a particularly high concentration of heavy REEs. These mineral deposits are often located along mid-ocean ridges, hydrothermal vent fields, and back-arc basins, where the heat from the Earth's mantle creates conditions conducive to the formation of these rich sulfide deposits (Milinovic et al., 2021; Vysetti, 2023).

Current estimates of the spatial extent of deep-sea mineral resources indicate that ferromanganese nodules cover approximately 38 million square kilometers of the ocean floor. Massive sulfide deposits, including both active and inactive sites, span around 3.2 million square kilometers, while polymetallic crusts found on seamounts are estimated to occupy about 1.7 million square kilometers (Petersen et al. 2016).

Most proposed methods for seabed mining involve deploying massive machinery. Seabed mining machinery typically consists of large, specialized equipment designed for underwater operations. A commonly used machine is the "crawler," resembling a heavy-duty, three-story tractor, which is equipped with high-powered mechanical arms or hydraulic systems. These machines traverse the ocean floor, scraping or drilling into the seabed to collect materials like polymetallic nodules or mineral-rich sediments. Some systems also use high-pressure water jets or vacuum technology to dislodge and collect the targeted resources, and they are typically controlled remotely due to the challenging environment at great depths. Once positioned on the ocean floor, the equipment would vacuum the uppermost 10 cm of the seabed, sending the material to a surface vessel via pipelines. Onboard the ship, the valuable minerals would be separated, while the remaining waste slurry would be returned to the ocean (Guo et al., 2023).

The extraction of REE from deep-sea environments poses significant environmental risks, particularly related to the disturbance of seabed sediments. These sediments often contain carbon dioxide (CO_2) that has been sequestered for millions of years. If disturbed by mining operations, this carbon could be released back into the atmosphere, exacerbating global warming. The disruption could also affect marine ecosystems, damaging biodiversity and food chains. Additionally, mining operations may introduce toxic substances and create sediment plumes that could further harm the delicate ocean floor habitats. Mining operations also play a significant role in the resuspension of previously deposited elements from sediments into the water column. This mechanical disturbance, whether through dredging, drilling, or sediment agitation, can release bound metals and metalloids, including REEs, that had previously settled and become relatively inert. Once resuspended, these elements may undergo changes in their chemical speciation due to shifts in pH, redox conditions, and interactions with dissolved organic matter, as noted previously. These transformations can significantly increase their bioavailability to aquatic organisms, enhancing the risk of uptake and bioaccumulation. Moreover, the mobilization of these contaminants may extend their ecological footprint, exposing previously unaffected areas and organisms to potential toxic effects. Another significant impact in this sense is the disturbance of seabed geology and topography. These changes may affect ocean currents, with downstream effects on coastal and marine climates. Additionally, the dispersal and migration patterns of numerous marine species could be severely impacted (Gausepohl et al., 2020). These cumulative impacts make deep-sea REE extraction a highly controversial practice.

Furthermore, mineral resources found on the seabed play a crucial role in the diversity of deep-sea ecosystems. For instance, polymetallic crusts in the deep sea provide hard surfaces for sessile animal communities like sponges to attach, as well as offering egg-laying sites for mobile species such as octopuses (Purser et al., 2016). Similarly, unique animal communities have adapted to survive the extreme temperatures and chemical conditions at hydrothermal vents, where massive sulfide deposits form as hot, mineral-rich fluids mix with cold seawater. These mineral deposits often act as "hotspots" of animal life, in contrast to the surrounding seafloor. A diverse range of microscopic organisms also thrives in these environments, responsible for the majority of the chemical cycling in these habitats, providing vital ecosystem services to deep-sea ecosystems (Thurber et al., 2014). For a detailed review on the impacts of deep-sea mining on microbial ecosystem services, see Orcutt et al. (2020).

Despite the ISA's regulatory framework, deep-sea mining remains a controversial topic due to the environmental risks and the uncertainty of long-term effects on marine ecosystems. Critics argue that current regulations may not go far enough to protect vulnerable marine life, while others worry that the economic interests in deep-sea mining may outpace efforts to develop effective environmental safeguards. In this sense, while the ocean holds promising potential for REE extraction, these endeavors come with considerable ecological risks. Therefore, any advancements in this field must be accompanied by comprehensive scientific research, thorough impact assessments, and robust regulatory measures to ensure that resource extraction does not come at the expense of long-term environmental preservation and ecosystem health (Van Dover, 2011; Da Ros et al., 2019; Sharma and Smith, 2019).

RARE EARTH ELEMENT INDUSTRIAL AND TECHNOLOGICAL APPLICATIONS

Rare Earth Elements have a wide range of industrial and technological applications, thanks to their unique chemical and physical properties. This places countries that consume and develop technologies using these elements at the forefront of industrial development (Sousa Filho and Serra, 2014).

Several important applications for these elements are noted (Ali and Katima, 2020). In the electronics sector, they are essential for the production of television screens, computers, mobile phones, silicon chips, and monitor displays, as well as long-life rechargeable batteries, camera lenses, LEDs, compact fluorescent lamps, baggage scanners, and even marine propulsion systems. Specifically, Eu and Y are vital for displays and LED technologies due to their phosphorescent properties, which also contribute to the development of advanced lighting systems, including compact fluorescent lamps and LED lights that are more energy-efficient than traditional options.

Their importance extends to manufacturing, where REEs contribute to the development of high-strength magnets, metal alloys, and stress gauges, while also being key components in ceramic pigments, colorants for glassware, chemical oxidizing agents, and polishing powders. Additionally, they enhance the creation of plastics, strengthen other metals, and improve the efficiency of automotive catalytic converters. In particular, cerium (Ce) and lanthanum (La) are widely used as catalysts in petroleum refining and in these converters, helping to reduce harmful emissions.

In the field of medical science, REEs are widely used in advanced imaging and diagnostic tools, including portable X-ray machines, X-ray tubes, and MRI contrast agents, as well as in nuclear medicine imaging and cancer treatment applications. Moreover, they are indispensable in genetic screening tests and are frequently employed in medical and dental lasers. Similarly, their role in technology is vast, encompassing applications such as lasers, optical glass, fiber optics, and masers, while also being critical for radar detection devices, nuclear fuel rods, and mercury vapor lamps. Additionally, they contribute to the development of highly reflective glass, computer memory, nuclear batteries, and high-temperature superconductors, making them fundamental to modern technological advancements.

Communication systems also benefit from REEs, particularly in fiber-optic signal amplification, which enhances energy efficiency and data transmission capabilities. In the renewable energy sector, they are indispensable in the production of permanent magnets for wind turbines, which eliminate the need for gearboxes, improve operational reliability—especially in offshore wind power generators—and facilitate the design of larger and more efficient power generation systems. This sector heavily relies on REEs such as neodymium (Nd), praseodymium (Pr), dysprosium (Dy), and terbium (Tb), which are used to manufacture the high-performance permanent magnets essential for wind turbines, electric vehicles, and advanced electronics.

Likewise, electric vehicles depend on REEs for the magnets in their electric motors, which play a key role in mitigating CO_2 emissions from the transport sector. Energy storage is another critical area where REEs make a significant impact, particularly in the development of nickel-metal hydride batteries used in electric and hybrid vehicles, as well as in rechargeable electronic devices. Their contribution extends to energy-efficient lighting, where they are essential for the production of fluorescent lamps and LEDs, offering a more sustainable alternative to traditional lighting solutions.

In the transport and energy sectors, REEs support hydrogen storage alloys for clean energy applications and are integral to ceramics used in hydrogen fuel cell vehicles and power generation. In fact, a typical hybrid automobile contains approximately 1 kg of rare earth elements, underscoring their importance in sustainable transportation. Beyond their role in energy and technology, REEs also contribute to environmental protection, particularly through their use in catalytic converters designed to reduce harmful emissions from exhaust gases.

Moreover, their applications extend to security and specialized technologies, as europium (Eu) is used in the authentication of euro bills to prevent counterfeiting, while holmium (Ho), which possesses the highest magnetic strength of any element, enables the creation of extremely powerful magnets that help reduce the weight of various motors. Additionally, the defense industry relies on several REEs for the production of advanced military technologies, including precision-guided munitions, missile systems, and communication devices, all of which are critical for modern defense strategies. With such a broad range of applications, REEs continue to be indispensable in driving innovation, sustainability, and technological progress across numerous industries. Table 2.2 summarizes the main current REE applications per sector and the most common REE employed in each.

Table 2.2 Main current REE applications

Sector	Application	Most Common REE
Electronics	– Television screens, computers, mobile phones, silicon chips, and monitor displays.	Eu, Y
	– Long-life rechargeable batteries, camera lenses, LEDs, compact fluorescent lamps, baggage scanners, marine propulsion systems.	Eu, Y
Manufacturing	– High-strength magnets, metal alloys, stress gauges, ceramic pigments, colorants for glassware, chemical oxidizing agents, polishing powders.	Ce, La
	– Plastics production, strengthening metals, automotive catalytic converters.	Ce, La
Medical Science	– Advanced imaging and diagnostic tools (portable X-ray machines, MRI contrast agents, X-ray tubes).	Eu, Y
	– Nuclear medicine imaging, cancer treatment, genetic screening tests, medical/dental lasers.	Eu, Y
Technology	– Lasers, optical glass, fiber optics, masers, radar detection devices, nuclear fuel rods, mercury vapor lamps, reflective glass, computer memory.	Eu, Y
	– Nuclear batteries, high-temperature superconductors.	Eu, Y
Communication Systems	– Fiber-optic signal amplification, energy-efficient data transmission.	Nd, Pr, Dy, Tb
Renewable Energy	– Permanent magnets for wind turbines (gearbox-free, improved reliability, larger, more efficient).	Nd, Pr, Dy, Tb
	– Electric vehicle motors (CO_2 emissions reduction), energy storage (Ni-MH batteries for electric/hybrid vehicles).	Nd, Pr, Dy, Tb
	– Energy-efficient lighting (fluorescent lamps, LEDs).	Eu, Y
Transport and Energy	– Hydrogen storage alloys for clean energy applications, ceramics in hydrogen fuel cell vehicles and power generation.	Nd, Pr, Dy, Tb
	– Hybrid vehicles containing approximately 1 kg of REEs.	Nd, Pr, Dy, Tb
Environmental Protection	– Catalytic converters for reducing harmful emissions from exhaust gases.	Ce, La
Security and Specialized Tech	– Europium for euro bill authentication to prevent counterfeiting.	Eu
	– Holmium for powerful magnets reducing motor weight.	Ho
Defense	– Advanced military technologies (precision-guided munitions, missile systems, communication devices).	Nd, Pr, Dy, Tb, Ho

ENVIRONMENTAL RARE EARTH ELEMENT IMPACTS

The mining and processing of REEs are resource-intensive activities that require significant energy and material inputs. These processes can also produce radioactive elements, and improper disposal of REE-containing products—such as their use in fertilizers—has been linked to diverse environmental impacts (Carpenter et al., 2015). For example, the refining of REEs often involves the use of various chemicals, which, if mishandled, can lead to environmental contamination. In an example of this, Bayan Obo, Mongolia, mining operations have resulted in a large toxic lake formed by extraction residues, highlighting the need for studies to evaluate the toxic effects of these elements on ecosystems and on human and animal health (Barry and Meehan, 2000). In another case, years of poorly regulated REE mining have led to the release of over 10 million tons of radioactive wastewater into a surrounding lake, rendering it highly toxic, as bastnaesite and monazite, the primary REE-bearing minerals, naturally contain radioactive elements like thorium and uranium (Mccord, 2021).

Although REEs are often considered less toxic compared to other metals (Sysolyatina and Olkova, 2022), their accumulation in the ecosystem and in organisms poses a potential risk to both environmental health and human consumption. However, ecotoxicological studies on REEs are still in their infancy, although the presence of these elements in the environment and humans has been documented (Herrmann et al., 2016). These elements are, in fact, known to accumulate in various environmental compartments, such as soil and water resources (Brioschi et al., 2013; Cidu et al., 2013) and can bioaccumulate and biomagnify along trophic chains (González et al., 2015). For example, certain REEs, such as Sm and Nd, can form soluble compounds like fluorides (e.g., SmF_3, NdF_3), chlorides (e.g., $SmCl_3$, $NdCl_3$), and sulfides (e.g., Sm_2S_3, Nd_2S_3). These forms are often more bioavailable than their oxide or phosphate counterparts because they dissolve more readily in water, especially under specific pH and redox conditions, which increase their bioavailability (Blaise et al., 2018).

Despite their use in agriculture as fertilizers, some REEs have been found to negatively affect plant growth. These effects include reduced root functionality, impaired nutrient absorption, and decreased seed germination (Thomas et al., 2014; Martinez et al., 2018; Tao et al., 2022). For example, Martinez et al. (2018) assessed the impact of REE on rice plant growth and whether iron (III) oxides on plant roots (iron oxide plaques) could reduce the toxic effects of REE. Hydroponic experiments were conducted with rice plants exposed to REE and iron salts (Fe(II) sulfate or Fe(II) chloride). Results showed significant reductions in root and plant height and biomass at REE concentrations of 0.5 mg L^{-1} and 1 mg L^{-1}. The addition of Fe(II) chloride or Fe(II) sulfate also caused negative growth effects. The presence of Fe(II) chloride seemed to mitigate some of the growth inhibition for light REE (La, Ce, Pr, Nd), while middle and heavy REE (Sm, Eu, Gd, and others) were largely bound to iron oxides. However, with Fe(II) sulfate, a toxic effect was observed for all REE concentrations, especially at higher levels. The study suggested that dissolved REE species, such as MREE-SO_4^+ and HREE-SO_4^+, might contribute to the growth inhibition. The negative growth effects were likely reduced due to REE sorption to iron (III) oxides on the root surface, potentially limiting their toxicity. Overall, the reported findings highlight that REE exposure is detrimental to rice growth, although iron oxide plaques may alleviate some of the toxicity. Another assessment examined the phytotoxicity of rare earth elements (REEs) lanthanum (La), yttrium (Y), and cerium (Ce) on three native Canadian plants (common milkweed, *Asclepias syriaca*; showy ticktrefoil, *Desmodium canadense*; and switchgrass, *Panicum virgatum*) and two common crop species (radish, *Raphanus sativus*; tomato, *Solanum lycopersicum*) (Thomas et al., 2014). Doses were added to the soil before sowing seeds, and the effects on germination and growth were measured. Germination effects were mainly observed with Ce at low pH, while growth inhibition was more pronounced, particularly for the native forbs, with 10–25% reductions in biomass for all REEs. Ce concentrations were higher in roots than in aboveground biomass, following the dose–response pattern. The authors, therefore, suggest that REEs are generally of limited toxicity and not considered extremely hazardous to the environment. However, in areas of potential REE contamination, their slow accumulation could pose environmental risks and trophic cascade effects, from primary producers such as plants to higher trophic levels.

The role of redox mechanisms in the biological effects of REEs, particularly terbium (Tb) and ytterbium (Yb), has been increasingly recognized. These elements can influence oxidative stress responses by acting as either antioxidants or prooxidants, depending on their concentration, chemical form, and the cellular environment (Pagano et al., 2015). At lower concentrations, some REEs may exhibit antioxidant properties, helping to neutralize reactive oxygen species (ROS) and protect cells. However, at higher concentrations or under certain physiological conditions, they may shift to prooxidant behavior, promoting ROS generation and oxidative damage to lipids, proteins, and DNA. This dual behavior is largely mediated through redox-active interactions involving cellular components like mitochondrial membranes or redox-sensitive enzymes.

Oxidative stress refers to an imbalance between free radicals (reactive oxygen species or ROS) and antioxidants in the body (Alkadi, 2020). Free radicals are unstable molecules that can damage cells, proteins, and DNA, leading to various health problems. Antioxidants counteract these free radicals, neutralizing them and preventing cellular damage (Pisoschi and Pop, 2015).

Oxidative stress occurs when the generation of reactive oxygen species (ROS) surpasses the body's antioxidant defense capacity. In this state, antioxidants play a crucial protective role by neutralizing harmful free radicals, thereby preserving cellular homeostasis. In contrast, prooxidants promote the formation of ROS, exacerbating oxidative damage. This imbalance can lead to the disruption of essential cellular processes and the structural degradation of key biomolecules such as DNA, proteins, and lipids, ultimately impairing organismal health. Despite growing experimental evidence of REE-induced oxidative stress, most studies are restricted to laboratory-based assessments using model organisms, while field-based investigations in natural ecosystems or wild fauna remain extremely limited. This gap hinders a comprehensive understanding of the ecological and health risks posed by REE contamination in real-world scenarios.

In one example, one study investigated the toxicity of La on the mussel *Mytilus galloprovincialis* exposed to various concentrations of La (0, 0.1, 1, 10 mg L^{-1}) for 28 days under controlled temperature and salinity conditions (Pinto et al., 2019). The results showed that La concentrations in mussels increased with higher exposure levels. La exposure led to metabolic changes, activation of antioxidant defenses (SOD, GPx), and biotransformation enzymes (GSTs). While oxidative stress was evident through a decrease in GSH/GSSG, oxidative damage was avoided, as indicated by lower levels of lipid peroxidation (LPO) and protein carbonylation (PC). Neurotoxicity was shown by the inhibition of acetylcholinesterase (AChE), and histopathological changes were observed in the gonads, gills, and digestive glands of mussels.

Another study evaluated the effects of different concentrations of Pr and Eu (0, 10, 20, 40, and 80 µg L^{-1}) on the mussel *Mytilus galloprovincialis* and their recovery from exposure to the highest concentration (Leite et al., 2024). Mussels accumulated both elements in a dose-dependent manner, with higher accumulation of Pr. Exposure to Pr increased the activity of antioxidant enzymes (superoxide dismutase and glutathione reductase) and biotransformation enzymes (carboxylesterases and glutathione S-transferases), but these defenses were insufficient to prevent cellular damage. All Eu concentrations caused cellular damage, despite an increase in enzyme activity. A histopathological analysis revealed that mussels could not recover from exposure to either element, with lower concentrations causing greater damage to digestive tubules. The authors conclude that exposure to Pr and Eu, even at low concentrations, leads to adverse mussel effects, potentially affecting their growth, reproductive capacity, and survival.

One assessment concerning PGMs compared the induction of heat shock protein 70 (hsp70) in zebra mussels (*Dreissena polymorpha*) exposed to soluble Pt, Pd, Rh, Cd and Pb salts (Singer et al., 2005). Mussels were sampled weekly for 10 weeks to measure metal concentrations and heat shock protein 70 (hsp70) levels. This protein is part of a broader family of heat shock proteins that are produced in response to environmental stressors, including temperature changes, heavy metals, pollutants, and oxidative damage a protein that helps other proteins fold correctly, preventing misfolding and aggregation, especially under stressful conditions such as heat, toxins, or oxidative stress. The highest metal uptake was observed for Cd, followed by Pt, Pb, and Pd, with Rh showing the lowest uptake. hsp70 levels increased over time in all exposed mussels, peaking at day 39 and then decreasing. The threshold metal concentrations for hsp70 induction followed the order: Rh $<$ Pd $\leq$ Pb $<$ Pt $<$ Cd. Mussels exposed to Pd showed the highest hsp70 induction, with a 25-fold increase, followed by Pt and Rh with 19-fold increases. Mussels exposed to Cd and Pb showed much lower increases, at 6- and 12-fold, respectively. The study concludes that Pt, Pd, and Rh strongly induce hsp70, with Pd having the greatest potential due to its low threshold and strongest effect compared to Cd and Pb.

Another study investigated the effects of Pt(IV) exposure on zebra mussels (*Dreissena polymorpha*) at various concentrations (0.1, 1, 10, 100, and 1,000 μg L^{-1}) over one and four days. Bioaccumulation was quantified and compared to alterations in biomarkers related to metal toxicity, including glutathione-S-transferase (GST) and catalase (CAT) activity, lipid peroxidation, and metallothionein (MT) induction. Platinum recovery in the exposure media ranged from 36% at 0.1 μg L^{-1} to 111% at 1,000 μg L^{-1}. Platinum concentrations in mussel tissues increased significantly, being 20–153 times higher than concentrations in the exposure media. Catalase activity was significantly elevated in mussels exposed to 0.1–100 μg L^{-1} Pt on Day 1, with a lowest effect concentration (LOC) of 0.1 μg L^{-1}. GST activity showed a similar, though less pronounced, trend. Lipid peroxidation was increased in mussels exposed to $\geq$0.1 μg L^{-1} Pt on Day 4. Pt bioaccumulation also correlated with an increase in Pt-MT.

RARE EARTH ELEMENT RECYCLING

Since REEs are non-renewable resources, their recovery and recycling are crucial for mitigating environmental impacts and ensuring economic sustainability (Xu and Peng, 2009). However, despite extensive lab-scale research on REE recycling, less than 1% of REEs are typically recycled, largely due to inefficient collection methods, technological challenges, and, particularly, a lack of incentives (Binnemans et al., 2013). Therefore, a significant improvement in REE recycling is essential, requiring the development of efficient, fully integrated recycling processes.

In this sense a study identified 1,835 individual landfills, primarily composed of waste rock, covering an area of about 100 km^2, with a total material REE quantity of 3.2 billion tons in South-Eastern Europe (Oliveira et al., 2021). Among these, 95 landfills contain approximately 1.6 billion tons of material, with an estimated total REE recovery potential of over 200 Kt. Red mud dams were also shown to contain high amounts, though their volume is smaller compared to the massive coal fly ash landfills.

Red mud dams, which store the highly alkaline by-product of bauxite refining in the aluminum industry, have also been shown to contain significant concentrations of REEs. While their total volume is generally smaller compared to the vast landfills generated by coal combustion, particularly coal fly ash (CFA), red mud still represents a considerable secondary source of REEs, due to their natural association with bauxite ores. Compared to CFA landfills, which are more widespread and voluminous due to global dependence on coal, red mud reservoirs offer a more concentrated but geographically limited source of REEs. Both are, however, increasingly being investigated as alternative resources for REE recovery to reduce reliance on primary mining and to support more sustainable circular economy approaches.

As a potential alternative, municipal solid waste incineration ash has recently been recognized for REE recovery, for example through a green modular treatment system with minimal secondary waste generation (Wen et al., 2023). One study, for example, successfully extracted over 80% of the total REEs at pH 2.0 and around 60% at pH 4.0 using citrate. Following this, an oxalate precipitation step selectively concentrated more than 98% of the extracted REEs, increasing their concentration by approximately 7–12 times compared to raw MSWIA. The waste byproducts were upcycled to synthesize zeolites, leading to an overall solid waste volume reduction of about 80% and heavy metal immobilization efficiency of approximately 75%, with negligible leaching, offering dual benefits of REE recovery and waste management with low chemical/energy input and a minimal environmental footprint.

Several studies have demonstrated the leaching of REEs from permanent magnets, fluorescent phosphors, catalysts, and other REE-containing wastes for recycling using acids such as sulfuric, hydrochloric, and nitric acids. For example, Diaz et al. (2016) and colleagues employed leaching with sulfuric or hydrochloric acid to extract REEs from the ferromagnetic fraction of e-wastes. In another study, Gergoric et al. (2018) used citric acid and acetic acid as leaching agents, and di-(2-ethylhexyl) phosphoric acid (D_2EHPA) was employed for preliminary solvent extraction tests. The highest leaching efficiencies were achieved with 1 mol L^{-1} citric acid, which leached nearly 100% of the REEs after 24 hours, and 1 mol L^{-1} acetic acid, which leached over 95% of the REEs. However, both Fe and Co, which are major impurities, were also co-leached, and no selectivity between the impurities and REEs was observed during the leaching process. In the solvent extraction tests using D_2EHPA in Solvent 70 on 1 mol L^{-1} leachates from both acetic acid and citric acid, a stronger affinity for Nd over Fe was observed, with better extraction properties in acetic acid leachate.

As an alternative to acids, some researchers have proposed using biologically generated leaching agents to extract REE, avoiding many of the environmental hazards associated with the corrosive chemicals used in conventional hydrometallurgy, as well as the high costs related to energy, reagents, and capital investments. Bioleaching leverages natural microbial activities, such as the production of organic acids or sulfuric acid. While bioleaching has long been employed for the commercial recovery of metals like gold and copper, recent studies have explored its application for REE recovery using bacteria and fungi. For example, Reed et al. (2016) used the gluconic acid-producing bacterium *Gluconobacter oxydans* to recover REEs from fluorescent phosphor wastes and spent petroleum fluid catalytic cracking catalysts. Hopfe et al. (2017) employed a combination of yeasts and bacteria in Kombucha culture to leach REEs from phosphor wastes. The ability to produce organic acids from a variety of substrates offers flexibility and allows for the production of low-cost leaching agents. Jin et al. (2019) demonstrated this by switching from glucose to agricultural wastes to feed *G. oxydans*, resulting in a mixture of organic acids equally effective for REE leaching, with environmental benefits and reduced costs. These advantages could make bioleaching a profitable and environmentally friendly option for secondary sources with dilute REE content, despite slower extraction rates compared to mineral acids.

Acid-free recovery of REEs has also been demonstrated as a viable approach, especially for materials like shredded e-waste, such as hard disk drives (HDDs) or magnet swarfs, using copper salt solutions without the need for energy-intensive steps such as pre-separation, pre-oxidation, or demagnetization, and generating no toxic waste (Prodius et al., 2019). In this method, NdFeB magnets undergo oxidative dissolution at room temperature, and the copper is precipitated and filtered for recycling. The solution containing REE^{3+} and Fe^{3+} is then treated with ammonium oxalate to precipitate REE-oxalate, which can be calcined to produce RE_2O_3, while the iron is recovered using standard methods. Another

Table 2.3 Main REE recycling alternatives

Study	Materials/Processes	REE Recovery/Concentration	Key Findings
Maât et al. (2016)	Acid-free REE Recovery (NdFeB Magnets)	95% Nd, Dy, Pr in magnetic fraction	Method using water and NaCl to extract REEs from NdFeB magnets with high purity, showing promise as a green chemistry approach.
Lister et al. (2016)	E-waste (Permanent Magnets, Fluorescent Phosphors)	High REE leaching with sulfuric and hydrochloric acids	Nearly 100% REEs extracted after 24 hours with sulfuric or hydrochloric acid, but co-leaching of Fe and Co occurred without selectivity.
Reed et al. (2016)	Bioleaching (Gluconobacter oxydans)	REEs recovered from phosphor wastes	Bacterial bioleaching with gluconic acid-producing bacteria showed effective REE recovery from fluorescent phosphor wastes and spent petroleum catalysts.
Hopfe et al. (2017)	Bioleaching (Kombucha Culture)	REEs leached from phosphor wastes	Yeasts and bacteria in a Kombucha culture leached REEs from phosphor wastes, showing flexibility and low-cost potential.
Gergoric et al. (2018)	E-waste (Permanent Magnets, Fluorescent Phosphors)	1 mol L^{-1} citric acid: 100% REEs leached	Citric acid and acetic acid achieved high leaching efficiencies for REEs, but impurities like Fe and Co were also leached.
Jin et al. (2019)	Bioleaching (Gluconobacter oxydans)	Effective REE leaching with agricultural wastes	Agricultural waste feeding for G. oxydans resulted in organic acids equally effective for REE leaching, offering environmental and cost benefits.
Prodius et al. (2019)	Acid-free E-Waste Recycling (Shredded HDDs, Magnet Swarf)	REE-oxalate precipitate (REE_2O_3 produced)	Copper salt solutions used for REE recovery from e-waste, with no toxic waste generation and no need for energy-intensive steps.
Wen et al. (2023)	Municipal Solid Waste Incineration Ash (MSWIA)	>80% of REEs extracted at pH 2.0, ~60% at pH 4.0	Citrate extraction and oxalate precipitation successfully extracted >98% of REEs, with a concentration increase of 7–12 times. Solid waste reduction of ~80% and heavy metal immobilization efficiency of ~75%.

acid-free approach using only water and sodium chloride (NaCl) (Maât et al., 2016). In this method, NdFeB magnets are heated to 250°C in a solution of water and NaCl to remove the nickel coating and oxidize the REE components (primarily neodymium) and iron into REE hydroxides and iron oxides. The two resulting powders can then be separated through magnetic separation. The metal portion of the magnetic fraction was found to be 95% Nd, Dy, and Pr, with iron being the primary impurity. Since the water can be reused and the NaCl concentration is low (<0.5 g/L), this method holds promise as a green chemistry approach. Table 2.3 summarizes the main REE recycling alternatives.

RARE EARTH ELEMENT LEGISLATION

Currently, there is only one piece of legislation specifically addressing the regulation of REEs, which applies to their concentrations in animal-derived food products. This legislative framework was established by China, a leading global player in both the production and regulation of REEs. The regulation stipulates that the total sum of all REE concentrations in food products should not exceed 0.7 mg kg^{-1} of the product (Jiang et al., 2012). This threshold was set with the aim of ensuring the safety of food consumed by the population, given the growing presence of REEs in various environmental and industrial contexts.

However, the regulation is currently limited in scope, as it only pertains to food products of animal origin and does not encompass other potential exposure pathways such as environmental contamination or the impact of REEs in other sectors. Furthermore, this piece of legislation has not yet been adopted on a global scale, and there is currently no widespread international consensus on the regulation of REEs in food products. As the use and demand for REEs continue to increase, it may be necessary to expand legislative frameworks to address the broader environmental and health implications of these elements.

RARE EARTH ELEMENTS IN AQUATIC ECOSYSTEMS

Several studies have evaluated REEs in aquatic ecosystems, mainly driven by the growing concern about their environmental implications. These studies generally aim to assess two primary aspects: the recycling potential of REEs, by assessing their distribution profiles in waterbodies, and their toxicity to exposed biota.

Concerning their distribution profiles in waterbodies, as REEs are often used in the production of high-tech devices, their accumulation in aquatic ecosystems through industrial discharges, mining runoff, and improper waste management, raises concerns about both environmental contamination and the possibility of resource recovery. Studies investigating this aspect focus on how these elements cycle through aquatic systems, how they are incorporated into sediments, and whether they can be re-mined or recovered from contaminated environments for reuse, potentially reducing the need for primary extraction. Regarding the latter, as these elements are increasingly found in various water bodies, concern about their impact on aquatic life, particularly species at different trophic levels has increased. Several studies have assessed the toxic effects of REEs on various aquatic organisms, including algae, crustaceans, mollusks, and fish. These studies evaluate critical parameters such as mortality rates, bioaccumulation, oxidative stress, and gene expression changes, providing insights into the potential risks REEs pose to individual organisms and broader ecosystem health.

In the context of REE recycling, one study examined the REE^{+3} series in seawater, noting a gradual enrichment from La to Lu, with Ce being strongly depleted due to its oxidation into the insoluble Ce^{+4} (Baar et al., 1991). The study observed that La and Gd exhibited modest enhancements relative to their neighboring REEs, possibly because of shifts in hydration around Gd. The complexation of REEs in seawater was primarily governed by carbonates, with an increasing influence as the atomic number rises, showing a strong dependence on oceanic pH (approximately 8.2). Adsorption onto particles was the main process responsible for the removal of REEs, with organic ligands showing lower formation constants for La and Gd. Carboxylic functional groups on particles were less effective at removing La and Gd, which likely contributed to their observed enrichment. These findings suggest that the vertical and lateral distributions of REEs in seawater resemble nutrient patterns, indicating their involvement in the organic geochemical cycle.

In the western Pacific, another study aimed to understand the mechanisms of REE enrichment by assessing seawater, porewater, and sediment samples (Deng et al., 2017). The results indicated enrichment of heavy REEs (HREEs) in both surface and deep seawater, while porewater showed HREE enrichment, low manganese (Mn) and iron (Fe) levels, and a negative Ce anomaly, all indicative of oxic conditions. The sediments exhibited high Mn/Al ratios and low uranium (U) concentrations, further supporting the presence of oxic conditions. Porewater profiles revealed enrichment of light REEs (LREEs) and middle REEs (MREEs) in the upper layers, while MREEs were depleted in deeper layers. The study suggested that porewater flux is a minor source of REEs to seawater, whereas REE enrichment in sediments is driven by extensive oxic conditions, low sedimentation rates, and strong adsorption, which contributes to the widespread presence of REE-rich sediments in the western Pacific basin.

Similarly, a study in the North Pacific measured dissolved (<0.04 μm) and total acid-soluble concentrations of REEs and yttrium in seawater from various depths in the western North Pacific near Japan using ICP mass spectrometry (Alibo and Nozaki, 1999). The acid-soluble particulate fraction was minimal, ranging from 2–5% for trivalent light and middle REEs and less than 1% for heavy REEs and yttrium. However, cerium (Ce) exhibited a significantly higher particulate fraction of 31%, consistent with its oxidation to the insoluble Ce(IV) state. Additionally, an elevated particulate fraction for all REEs was observed within approximately 80 m above the seafloor, attributed to the presence of flocculated resuspended particles. Vertical REE(III) profiles displayed smoothly increasing convex curves with depth, which were consistent with previously reported trends. The dissolved Ce concentrations decreased from around 6 pmol/kg near the surface to a minimum of 2.5 pmol kg^{-1} at 400 m depth, where the North Pacific Intermediate Water penetrates, before stabilizing at 4 pmol kg^{-1} below 800 m. In contrast, particulate Ce concentrations increased from the surface to 200 m, indicating the oxidation of Ce(III) to Ce(IV) and its subsequent scavenging in the upper water column. However, no evidence suggested that Ce oxidation continued in the deep sea. Shale-normalized patterns of dissolved REEs, examined using composite shale datasets from the literature, revealed distinct seawater signatures, including a well-documented negative Ce anomaly, a positive La anomaly, and a slight positive Gd anomaly. These anomalies varied systematically with depth, with rapid changes occurring in the upper several hundred meters, indicating that REE distributions are primarily influenced by ocean circulation and biogeochemical processes.

In the Sargasso Sea, a study conducted by Sholkovitz et al. (1994) investigated the distribution of REEs in suspended particles using sequential digestion with acetic acid, mild HCl/HNO_3, and HF/HNO/ HCl. The acetic acid and mild HCl/HNO_3 treatments were used to remove surface coatings, including organic matter and Mn oxides, while the latter two dissolved detrital minerals. Unlike the crust-like REE composition of detrital phases, surface coatings exhibited extensive fractionation compared to both filtered seawater and crust. These coatings play a crucial role in the removal and fractionation of REEs from seawater, significantly influencing marine geochemical cycles. Compared to seawater, surface coatings were enriched tenfold in trivalent REEs from Lu to La, exhibiting large positive Ce anomalies. These anomalies shifted from negative (seawater-like) in the upper 100 m to strongly positive at greater depths, a transformation likely due to the *in situ* oxidation of dissolved Ce(III) to particulate Ce(IV) and the preferential removal of LREE(III)s over HREE(III)s. This REE fractionation was aligned with particle/solution models and linked to Mn oxide particle formation through the oxidation of dissolved Mn(II) in the upper 200 m. While LREEs were preferentially removed in surface waters, their remineralization at depth led to their preferential release.

A study by Rétif et al. (2024a) compared the distribution of REEs in the abiotic compartments of water (dissolved phase <0.45 μm), suspended particulate matter (SPM), and sediment in the Loire and Seine estuaries, in France. A total of eight sites in the Loire and six in the Seine were investigated, along with five additional offshore sites in the Loire. Total REE concentrations were higher in the Loire for the dissolved phase (93.5 ± 63.3 vs. 87.7 ± 16.2 ng/L), SPM (173.9 ± 18.3 vs. 114.0 ± 17.8 mg kg^{-1} dw), and sediments (198.2 ± 27.9 vs. 73.2 ± 27.4 mg kg^{-1} dw), attributed to higher geogenic inputs. Individual REE contributions and normalization revealed heavy REE enrichment and positive Gd anomalies in the dissolved phase of both estuaries, whereas REE distributions in SPM and sediments reflected the natural abundance of REE classes. The calculated Gd anomalies in the dissolved phase were higher in the Seine (9.7 ± 3.4) than in the Loire (3.0 ± 0.8), corresponding to 88.3 ± 5.1% and 64.4 ± 11.1% of anthropogenic Gd, respectively. This suggests greater contamination in the Seine estuary, likely due to the larger population and higher medical use of Gd-based contrast agents. Offshore sites in the Loire exhibited lower total REE concentrations (55.8 ± 5.8 ng L^{-1}, 26.7 ± 38.2 mg kg^{-1} dw, and 100.1 ± 11.7 mg kg^{-1} dw for the dissolved phase, SPM, and sediments, respectively) and lower Gd anomalies (1.2 ± 0.2), with only 13.3 ± 3.9% of anthropogenic Gd, confirming contamination from the watershed.

Mussels, particularly their soft tissues and shells, are increasingly used as bioarchives for REE in environmental science and geology context (Briant et al., 2021). One study in this sense examined the bioavailability of anthropogenic REEs in Germany's Rhine and Weser Rivers, both of which receive significant REE inputs from human activities, *i.e.*, both rivers contain anthropogenic Gd from MRI contrast agents, and the Rhine also carries anthropogenic La and Sm from an industrial source (Merschel and Bau, 2015). To assess bioavailability, the aragonitic shells of the freshwater bivalve *Corbicula fluminea* were analyzed. The findings indicated that geogenic REE concentrations varied by sampling site and mussel age. Despite high Gd anomalies in the dissolved river water, no anomalous Gd enrichment was found in the shells, suggesting geogenic and anthropogenic Gd differ in speciation, with only the geogenic form being incorporated. In contrast, shells from sites downstream of the industrial source showed clear La and Sm anomalies, indicating their bioavailability. Table 2.4 summarizes the main findings of these studies concerning REE distribution profiles in waterbodies and their different environmental compartments.

Table 2.4 Summary of the main findings of studies on the REE distribution profiles in waterbodies and their different environmental compartments

Study	Location	Matrix	Key Observations
Baar et al. (1991)	Western Pacific Ocean	Seawater	Ce is strongly depleted due to oxidation to Ce(IV). LREE and HREE exhibit different enrichments. Vertical distributions resemble nutrient patterns, suggesting organic geochemical involvement.
Sholkovitz et al. (1994)	Sargasso Sea	Suspended particles	Positive Ce anomalies in particles, with significant REE fractionation. $Ce^{3}+$ oxidized to Ce^{4+}. LREE removal greater than HREE in surface waters.
Alibo and Nozaki (1999)	Western North Pacific near Japan	Seawater	Ce oxidation to Ce(IV) observed in vertical profiles. Ce concentrations decrease with depth. Particulate Ce increases at shallow depths. REE patterns showed a negative Ce anomaly and positive La and Gd anomalies.
Merschel and Bau (2015)	Rhine and Weser Rivers, Germany	Freshwater bivalves (bioavailability proxies)	Gd anomalies in river water but no Gd enrichment in mussels, indicating different speciation. Bioavailability of REEs depended on their geogenic or anthropogenic form.
Deng et al. (2017)	Western Pacific Ocean	Seawater, porewater, and sediments	HREE enrichment in seawater and porewater. Sediments showed high Mn/Al ratios and low U concentrations, supporting oxic conditions. Enrichment patterns in sediments likely due to low sedimentation and strong adsorption.
Rétif et al. (2024a)	Loire and Seine Estuaries, France	Water, suspended particulate matter and sediments	REE concentrations higher in Loire than Seine. Positive Gd anomalies in dissolved phases. Gd anomalies in the Seine were higher, indicating more anthropogenic influence. Offshore sites had lower REE concentrations, confirming watershed contamination.

Regarding toxicity to exposed aquatic biota, several studies have noted moderate to high toxicity of several REEs. Most assessments, however, have been conducted in laboratory settings, and have focused on individual elements rather than their natural mixtures, although some studies are available in this sense. In addition, most studies have been conducted on model organisms, such as algae, microcrustaceans and some bivalves, with scarce assessments on fish and other taxa available. In this sense, one recent review article analyzed 38 papers on REE ecotoxicity, reporting that most studies have focused on freshwater ecosystems, highlighting a gap in research on marine environments, particularly concerning marine vertebrates (Revel et al., 2025). The results also indicated that heavy REEs tend to be more toxic to aquatic organisms than light REEs, and several key factors were identified as crucial in bioassays of REE toxicity, including the complexation of REEs with ions like phosphates and carbonates. The presence of carbonates was noted as decreasing the bioavailability of REE species in water, potentially reducing toxicity, though competition between REE^{3+} and Ca^{2+} for the same binding sites in organisms may also contribute to this effect. Selected studies, both laboratory and field-based, from lower to higher taxa are discussed below on model organisms and free-ranging biota in diverse aquatic systems.

Several studies have been conducted on REE toxicity, mainly in laboratory exposure settings. In this context, one study investigated the biological toxicity of lanthanides on marine monocellular algae, examining the relationship between lanthanide abundance in seawater and their toxic effects on *Skeletonema costatum* (Tai et al., 2010). The results showed that all individual lanthanides had similar toxic effects on the algae. High concentrations of lanthanides (29.04 ± 0.61 µmol L^{-1}) resulted in a 50% algae growth reduction after 96 hours compared to the controls (0 µmol L^{-1}). The biological toxicity of 13 lanthanides was found to be unrelated to their natural abundance. A mixed solution containing equal concentrations of each lanthanide had the same inhibitory effect on algae cells as each individual lanthanide at the same total concentration. This phenomenon is unique compared to other element groups in the periodic table, suggesting that monocellular organisms might not be able to sufficiently differentiate between the chemically similar lanthanide elements. Another study, also on algae, focused on the environmental risk assessment of La and Ce, by evaluating their toxicity in relation to metal speciation in solution in the freshwater green alga *Chlorella fusca* (Aharchaou et al., 2020). The research found that the calculated free ion concentration of both La and Ce was a good predictor of their toxicity. When organic ligands were present or absent, the half-maximal effective concentrations based on free ion concentrations were similar. The median effective concentration values for La^{3+} and Ce^{3+}

after 120 h of exposure were 0.47–0.61 μmol L^{-1}, showing identical toxicity thresholds for both elements. The inhibition of algal growth was directly correlated with the intracellular lanthanide concentrations, irrespective of the ligand used. Additionally, an increased ambient calcium concentration protected algae by reducing the internalization of lanthanides into the cells. The findings suggest that, at a constant pH of 5.5, the accumulation and toxicity of REEs are linked to both free ion concentration and ambient calcium levels.

Another study investigated REE toxicity using *Hydra attenuata* as an animal model, a small freshwater hydroid species of ecological relevance (Blaise et al., 2018). Lethal and sublethal toxicity data for 11 REEs showed LC50 values ranging from 0.21 to 0.77 mg L^{-1} and EC50 values from 0.02 to 0.27 mg L^{-1}, confirming that *Hydra attenuata* is highly sensitive to REE exposure at environmentally relevant concentrations. The study also revealed that the toxicity of REEs decreases with increasing atomic number and decreasing ionic radius. Compared to other taxonomic groups, *Hydra attenuata* displayed one of the highest sensitivities to REEs, similar to other invertebrate species like *Daphnia magna*, *Ceriodaphnia dubia*, and *Hyalella azteca*. The authors indicate that the findings suggest that members of this community are likely more vulnerable to REE discharges in aquatic environments. Given *Hydra's* sensitivity, it was proposed as a valuable species for inclusion in toxicity testing to assess liquid samples suspected of containing REEs.

In another assessment, the toxicity of La and Ce were evaluated in sea urchin (*Paracentrotus lividus*) embryos and sperm (Oral et al., 2010). Embryos were exposed for 72 h to Ce^{+4} or La^{+3} in seawater at concentrations ranging from 10^{-8} to 10^{-5} mol L^{-1}. A cytogenetic analysis assessed mitotic activity and mitotic aberrations in 5-hour post-fertilization embryos, while an embryological analysis determined developmental anomalies and mortality rates. Additionally, sperm were exposed to Ce(IV) or La(III) (10^{-8}–10^{-5} mol L^{-1}) for 1 h before fertilization, and subsequent embryo development was monitored. Exposure to 10^{-5} mol L^{-1} Ce(IV) led to 100% embryonic mortality, whereas 10^{-5} mol L^{-1} La(III) caused 100% developmental defects without mortality. Ce(IV)-exposed embryos (10^{-7} to 3×10^{-6} mol L^{-1}) showed significant mitotic inhibition and an increase in mitotic aberrations, effects not observed with La(III). Sperm exposure to the highest tested concentration (10^{-5} mol L^{-1}) reduced fertilization success for both Ce(IV) and La(III). However, only offspring from Ce(IV)-exposed sperm exhibited a significant concentration-related increase in developmental defects. The authors indicate that these findings suggest that REEs, particularly Ce^{+4}, may pose significant risks to marine biota, highlighting the need for further studies on a broader range of REEs and their environmental impacts.

In another study, the acute and long-term toxicities of selected lanthanide nitrates (La, Ce, Pr, Nd, and Gd) to freshwater crustaceans *Daphnia magna, Thamnocephalus platyurus,* and *Heterocypris incongruens* were assessed (Blinova et al., 2018). The findings showed that acute toxicity data for lanthanides are unreliable due to methodological issues and should not be used for risk assessments, and that a 21-day *D. magna* reproduction test revealed high toxicity, with parent mortality being a more sensitive endpoint than reproduction. Furthermore long-term LC_{50} values for the investigated lanthanides ranged from 0.3 to 0.5 mg L^{-1}, with no statistically significant differences between individual lanthanides. The study concluded that environmental risk assessments of lanthanides should rely on long-term toxicity tests, and that lanthanides may be considered a uniform group of elements with an additive mode of action until further research clarifies potential differences in their ecotoxicity mechanisms.

One assessment evaluated the effects of mixtures of three lanthanides, namely Ce, Gd and Lu, and on seven aquatic species: *A. fischeri, R. subcapitata, C. vulgaris, B. calyciflorus, H. incongruens, D. magna,* and *D. rerio* (Romero-Freire et al., 2019b). Concentrations decreased over time in the test media, with sedimentation starting early and reducing the available elements. Toxic effects were observed in five of the seven species, particularly in the unicellular *A. fischeri* and organisms at lower trophic levels, such as *R. subcapitata* and *B. calyciflorus.* Synergistic effects were found in bacteria and algae tests, while antagonistic effects were seen in rotifers. The study highlights the complexity of predicting toxicity in multi-element exposures, noting that interactions with other compounds or nutrient removal could mask effects. The varying toxic responses among species also suggest that lanthanide interactions should not be overlooked, and that further research on their mixtures is necessary to understand their toxicity in natural ecosystems. Another study tested chloride salts of cerium ($CeCl_3 \cdot 7H_2O$), gadolinium ($GdCl_3 \cdot 6H_2O$), and lutetium ($LuCl_3 \cdot 6H_2O$) were tested on *A. fischeri, Pseudokirchneriella subcapitata, D. magna, H. incongruens, B. calyciflorus,* and *Hydra attenuata* (González et al., 2015), reporting the sensitivity of *A. fischeri* and *P. subcapitata* increased with the atomic number of the lanthanides.

Yet another study evaluated the toxicity of seven REEs (Y, La, Ce, Nd, Sa, Eu, Gd) on early life stages of two other sea urchin species, *Sphaerechinus granularis* and *Arbacia lixula* (Trifuoggi et al., 2017), tested as chloride salts at concentrations ranging from 10^{-7} to 10^{-4} mol L^{-1}, with endpoints including developmental defects, cytogenetic anomalies, fertilization success, and offspring damage following sperm exposure. The results revealed distinct toxicity patterns for each REE, varying according to the species and treatment protocol, indicating toxicity scaling among the different REEs. *S. granularis* showed significantly higher sensitivity compared to *A. lixula* and *P. lividus*. The study provides clear evidence of species-specific toxicity to REEs, with implications for assessing species susceptibility along polluted coasts.

In another study, the toxicity of three representative REEs, Nd, Gd, and Yb, on three freshwater species representing different trophic levels were assessed, namely algae, crustaceans, and fish (Lachaux et al., 2022). The toxicity of REEs was also tested in mixtures. Additionally, dissolved organic matter (DOM) was added to the test medium since DOM is ubiquitous in freshwater systems and influences REE speciation. The results showed that DOM significantly reduced REE bioaccumulation and toxicity, likely by forming non-bioavailable REE-DOM complexes. Among the tested species, algae were the most sensitive to REEs. Although there were slight differences between Nd, Gd, and Yb in terms of behavior and bioaccumulation, the three REEs showed comparable toxicity and additive effects in mixtures across all organisms. Therefore, the study considered REEs as a uniform group.

Regarding fish models, one study investigated the lethal and sublethal toxicity of REEs in juvenile rainbow trout (*Oncorhynchus mykiss*) (Dubé et al., 2019). The fish were exposed to increasing concentrations (0.064, 0.32, 1.6, 8, and 40 mg L^{-1}) of seven REEs (cerium, erbium, gadolinium, lanthanum, neodymium, samarium, and yttrium) for 96 hours at 15°C. Mortality rates were measured, and in surviving fish, gene expression changes in the liver were tracked across 10 target genes related to oxidative stress, DNA repair, tissue growth/proliferation, protein chaperoning, xenobiotic biotransformation, and ammonia metabolism. Results revealed that Y, Sm, Er, and Gd formed a distinct group based on toxicity (mortality) and gene expression changes. A significant correlation was found between electronegativity and lethal concentration (LC_{50}). Gene expression changes occurred at concentrations around 120 times lower than the LC_{50}, with key transcripts related to heat shock proteins, DNA repair (growth arrest DNA damage), and CYP1A1 involved in xenobiotic metabolism. The study concluded that the more electronegative REEs were more toxic to juvenile trout and caused sublethal effects at concentrations two orders of magnitude lower than lethal doses. The authors indicate that REE toxicity depends on their chemical properties and involves specific biological pathways at the gene expression level. Another study on juveniles of the same species exposed fish to a mixture of the five most abundant REEs for 96 hours, at concentrations resembling those found in lakes contaminated by mining activities (Hanana et al., 2021). The 1x mixture contained cerium (Ce, 280 μg L^{-1}), lanthanum (La, 140 μg L^{-1}), neodymium (Nd, 120 μg L^{-1}), praseodymium (Pr, 28 μg L^{-1}), and samarium (Sm, 23 μg L^{-1}). The expression of 14 genes related to oxidative stress, DNA repair, tissue growth/proliferation, protein chaperoning, xenobiotic biotransformation, and ammonia metabolism in the liver were investigated. Additionally, DNA damage, oxidative stress (lipid peroxidation or LPO), inflammation (cyclooxygenase or COX activity), detoxification (glutathione-S-transferase activity or GST), and labile zinc levels in gills were also assessed. The results showed that genes involved in oxidative stress, such as catalase (*cat*), heat shock proteins 70 (*hsp70*), and glutamate dehydrogenase (*glud*), were upregulated, while glutathione S-transferase (*gst*) and metallothionein (*mt*) gene expressions were downregulated. The mixture exhibited genotoxic effects and increased labile zinc levels in the gills of exposed trout. These changes occurred at concentrations 600 times lower than the LC50 for the mixture, indicating effects even below the 1X concentration. Based on principal component analysis and concentration-dependent responses, the most significant sublethal effects included DNA strand breaks (genotoxicity), labile zinc, cat, gst, hsp70, sparc, mt, and glud. The authors indicate that the findings suggest that similar effects could occur in environments impacted by mining activities.

One study in another model organism, the zebrafish (*Danio rerio*), investigated the toxic effects of La and Gd through toxicity experiments and transcriptome analyses (Kang et al., 2022). The toxicity experiments showed that both La and Gd exhibited similar lethality, with half-lethal concentrations (LC_{50}) at micromolar levels and an additive toxicity effect. Differential gene expression screening and functional group enrichment analysis revealed that both REEs may impact the growth and development of zebrafish by interfering with various biological molecules. The two REEs notably affected metabolic pathways of exogenous or endogenous substances, including glutathione sulfotransferase and acetaldehyde dehydrogenase. In addition, key biological processes such as DNA replication, the insulin signaling pathway, and the p53 signaling pathway were significantly enriched. Overall, while La and Gd exhibited differences in toxicity intensity, their mechanisms and modes of action shared many similarities. The concentrations used were comparable to REE residues found in highly contaminated sites, effectively mimicking the ecotoxicological effects of REEs at environmentally relevant concentrations. Table 2.5 summarizes the main findings of these studies for laboratory assessments on REE toxicity.

Concerning field studies, one assessment measured REE concentrations and stable isotope ratios ($\delta^{15}N$, $\partial^{13}C$) in biota from marine, freshwater, and terrestrial ecosystems in the eastern Canadian Arctic (MacMillan et al., 2017). The results demonstrated that REEs follow a coherent bioaccumulation pattern, though some redox-sensitive elements (Ce, Eu) had anomalies. The highest REE concentrations were found at

Table 2.5 Summary of the main findings of studies concerning laboratory-based assessments on REE toxicity

Study	Tested Species	Tested REEs	Key Findings
Tai et al. (2010)	*Skeletonema costatum* (diatom)	La, Ce, Pr, Nd, Sm, Eu, Gd, Tb, Dy, Ho, Er, Tm, Yb	Toxicity of individual lanthanides similar. High concentrations (29.04 ± 0.61 μmol L^{-1}) caused 50% growth reduction. Toxicity unrelated to natural abundance of lanthanides.
Oral et al. (2010)	*Paracentrotus lividus* (sea urchin)	La^{3+}, Ce^{4+}	100% mortality at 10^{-5} mol L^{-1} Ce^{4+} and 100% developmental defects at 10^{-5} mol L^{-1} La^{3+}. Ce^{4+} led to mitotic inhibition and significant developmental defects.
Trifuoggi et al. (2017)	*Sphaerechinus granularis, Arbacia lixula* (sea urchins)	Y, La, Ce, Nd, Sm, Eu, Gd	Toxicity varied by REE and species. S. granularis more sensitive than other species, with implications for assessing species susceptibility in polluted coasts.
Blinova et al. (2018)	*Daphnia magna, Thamnocephalus platyurus, Heterocypris incongruens* (crustaceans)	La, Ce, Pr, Nd, Gd	Acute toxicity unreliable for risk assessment; 21-day D. magna test showed high long-term toxicity. LC50 values ranged from 0.3 to 0.5 mg L^{-1}.
Blaise et al. (2018)	*Hydra attenuata*	La, Ce, Pr, Nd, Sm, Eu, Gd, Tb, Dy, Ho, Er, Tm, Yb	LC50 values ranged from 0.21 to 0.77 mg L^{-1}. Toxicity decreases with increasing atomic number and decreasing ionic radius. Hydra showed high sensitivity to REEs, suggesting vulnerability in aquatic environments.
Dubé et al. (2019)	*Oncorhynchus mykiss* (juvenile rainbow trout)	Ce, Er, Gd, La, Nd, Sm, Y	Y, Sm, Er, and Gd formed a distinct toxicity group. Significant correlation between electronegativity and LC50. Sublethal effects seen at concentrations 120x lower than LC50.
González et al. (2015)	*Aliivibrio fischeri, Pseudokirchneriella subcapitata, Daphnia magna, Heterocypris incongruens, Brachionus calyciflorus, Hydra attenuata*	Ce, Gd, Lu	Sensitivity increased with atomic number. Toxicity was dependent on the REE tested.

Table 2.5 (Contd.)

Table 2.5 (*Contd.*) Summary of the main findings of studies concerning laboratory-based assessments on REE toxicity

Study	Tested Species	Tested REEs	Key Findings
Romero-Freire et al. (2019b)	*Aliivibrio fischeri, Raphidocelis subcapitata, Chlorella vulgaris, Brachionus calyciflorus, Heterocypris incongruens, Daphnia magna, Danio rerio*	Ce, Gd, Lu	Synergistic effects in bacteria and algae, antagonistic effects in rotifers. Toxicity varied across species. Interaction effects complicate toxicity predictions.
Aharchaou et al. (2020)	*Chlorella fusca*	La, Ce	Toxicity predicted by free ion concentration of La^{3+} and Ce^{3+}. Half-maximal effective concentration values for La^{3+} and Ce^{3+} were similar. Increased calcium concentration protected algae.
Hanana et al. (2021)	*Oncorhynchus mykiss* (juvenile rainbow trout)	Ce, La, Nd, Pr, Sm	Gene expression changes at concentrations 600x lower than LC50. Significant sublethal effects like genotoxicity and labile zinc increase.
Kang et al. (2022)	*Danio rerio* (zebrafish)	La, Gd	La and Gd exhibited similar lethality, with additive toxicity. Gene expression changes affected various metabolic and DNA replication pathways.
Lachaux et al. (2022)	*Algae, crustaceans, fish* (general species)	Nd, Gd, Yb	Algae were the most sensitive to REEs. DOM reduced bioaccumulation and toxicity. REEs exhibited comparable toxicity, showing additive effects in mixtures.

low trophic levels, especially in vegetation and aquatic invertebrates. Terrestrial herbivores, ringed seals, and fish had low REE levels in muscle tissue, but accumulation was much higher in liver tissues. Age- and length-dependent REE accumulation suggested faster uptake than elimination in some species. Overall, REE bioaccumulation was species- and tissue-specific, with limited biomagnification.

One study assessed REE concentrations in *Mytilus galloprovincialis* from six locations along the Portuguese coast, analyzing seasonal variations and potential links to local ecosystem characteristics (Figueiredo et al., 2022a). Mussels from Porto Brandão (south bank of the Tagus estuary) exhibited the highest REE concentrations, while those from Aljezur (southernmost site) had the lowest in both autumn and spring. Overall, total REE concentrations (ΣREE) were higher in spring. Light REEs (LREEs) were more enriched than heavy REEs (HREEs), and negative anomalies for Ce and Eu were observed.

Another field-based assessment measured the concentrations of REEs in ten freshwater fish species from a reservoir in Washington State, USA (Mayfield and Fairbrother, 2015). The distribution of REEs was examined in both fillet and whole-body tissues across three different size classes. The total concentrations (ΣREE) ranged from 0.014 to 3.0 mg kg^{-1} dry weight, with an average concentration of 0.243 mg kg^{-1} dry weight. The findings indicated that REEs accumulated more in organs, viscera, and bones compared to muscle (fillet) tissues. Fish species that fed on benthic organisms (exposed to sediments) exhibited higher REE concentrations than pelagic omnivorous or piscivorous species. Additionally, for certain species like the largescale and longnose suckers, smallmouth bass, and walleye, REE concentrations decreased with increasing age, total length, or weight. The observed concentration patterns were consistent with natural conditions and indicated no significant anthropogenic sources of REEs in this system.

Yet another study quantified the total concentrations of rare earth elements (ΣREE) in 251 samples from 10 common species of freshwater and marine fish collected from seventeen cities in Shandong, China (Yang et al., 2016). The ΣREE concentrations in freshwater fish ranged from 34.0 to 37.9 ng g^{-1} (wet weight), while in marine fish, they ranged from 12.7 to 37.6 ng g^{-1}. The ratio of light rare earth elements (LREE) to heavy rare earth elements (HREE) was 13.7:1 in freshwater fish and 10:1 in marine fish, indicating that freshwater fish tend to have higher REE concentrations than marine fish, and that LREEs exert greater biological effects than HREEs. The REE distribution patterns in the studied fish were similar to those found in coastal sediments, supporting the "abundance law" of element distribution, referring to a principle that suggests the concentration patterns of elements in biological organisms (such as fish) tend to mirror their relative concentrations in the environment, particularly in sediments. Health risk assessments showed that the estimated daily intakes (EDIs) of REEs from fish were significantly lower than the acceptable daily intakes (ADIs), suggesting that consuming these fish poses little risk to human health.

Another study was conducted in the northern coastal region of the South China Sea and assessed 15 REEs in 14 marine fish species (Wang et al., 2022). Total REE concentrations (ΣREEs) ranged from 1.02 to 178.55 $\mu g\ kg^{-1}$ (wet weight), with light REEs (LREEs) dominating over heavy REEs (HREEs) at an average ratio of 13.04. Significant bioaccumulation of LREEs was observed, along with Ce, Gd, Tb, and Dy anomalies in most species. Based on estimated daily intake (EDI) assessments, the authors indicate that consuming the local fish species poses negligible health risks to consumers. Table 2.6 summarizes the main findings reported for field-based studies.

Table 2.6 Summary of the main findings of studies concerning field-based assessments on REE concentrations and bioaccumulation trends

Study	Location	Species	REE Concentrations and Trends
Mayfield and Fairbrother (2015)	Washington State, USA	10 freshwater fish species	ΣREE: 0.014–3.0 $mg\ kg^{-1}$ dry weight
Yang et al. (2016)	Shandong, China	Freshwater and marine fish species	Freshwater fish: 34.0–37.9 $ng\ g^{-1}$ wet weight; Marine fish: 12.7–37.6 $ng\ g^{-1}$ wet weight
MacMillan et al. (2017)	Eastern Canadian Arctic	Terrestrial herbivores, ringed seals, fish, vegetation, aquatic invertebrates	Highest concentrations in vegetation and aquatic invertebrates (specific concentrations not listed)
Wang et al. (2022)	South China Sea	14 marine fish species	ΣREE: 1.02–178.55 $\mu g\ kg^{-1}$ wet weight
Figueiredo et al. (2022a)	Portuguese Coast	*Mytilus galloprovincialis* (mussels)	Highest REE concentrations in Porto Brandão (specific concentrations not listed).

In a rare example of a trophic food web REE assessment, one study focused on the distribution of REE across a food web in the Seine estuary, an area affected by increasing anthropogenic use of REEs (Rétif et al., 2024b). The research included species from five phyla, representing different trophic levels, to assess how REEs are transferred and accumulated in aquatic organisms. The total concentrations of REEs were found to be higher in primary producers and lower trophic level organisms, such as algae, mollusks, crustaceans, and annelids, ranging from 0.14 to 156 µg g^{-1} dry weight. In contrast, concentrations were much lower in vertebrates, ranging from 0.03 to 0.15 µg g^{-1} dry weight. The trend of higher REE concentrations in lower trophic levels correlated with δ15N values, indicating a trophic dilution effect, where higher trophic level organisms accumulate fewer REEs due to the decreased concentration in their prey. The study also revealed that light REEs (LREEs) were more prevalent than medium and heavy REEs across all species studied. Furthermore, certain REEs like Eu, Gd, Tb, and Lu showed positive anomalies, especially in vertebrates. These anomalies could be attributed to species-specific bioaccumulation and detoxification processes or possibly due to anthropogenic inputs of these elements into the ecosystem. Bioaccumulation factors (BAF) and biota-sediment accumulation factors (BSAF) calculated for the species indicated that there was a significant partitioning of REEs within the organisms compared to the dissolved phase, suggesting a limited transfer of REEs from sediments to organisms. The findings highlight the role of biota in concentrating REEs from the surrounding environment, but also shows that REE uptake from sediment is not a major pathway in this estuary system.

One study assessed REE distribution in 14 Canadian temperate lakes unaffected by mining to determine their natural ecosystem fate, providing essential baseline data on REE concentrations and trophic dynamics (Amyot et al., 2017). Results showed strong correlations between individual and total REE concentrations (ΣREE) across different components of lake food webs. Dissolved organic carbon, dissolved oxygen, and sediment ΣREE were identified as key drivers of aqueous REE levels. Median bioaccumulation factors (log10) ranged from 1.3 in fish muscle to 4.4 in nonpredatory invertebrates, with higher accumulation in lower trophic levels. ΣREE in fish, macroinvertebrates, and zooplankton declined with increasing trophic position, as indicated by nitrogen isotopes ($\delta^{15}N$), demonstrating trophic dilution. Low REE concentrations in fish muscle suggest minimal risk to human consumers in non-mining areas. However, piscivorous birds and mammals may accumulate higher REE levels from whole fish consumption.

Another study analyzed the concentrations of 15 naturally occurring REEs in seven common species from Maluan Bay, China, revealing species- and element-specific bioaccumulation patterns, even among organisms within the same taxa or phylum (Wang et al., 2019). However, nitrogen isotope signatures ($\delta^{15}N$) indicated limited potential for biomagnification. Chondrite-normalized REE distribution patterns were similar across samples, suggesting a common REE source, with occasional anomalies for redox-sensitive Ce and Eu. Light REEs (La to Eu) were significantly more abundant than heavy REEs (Gd to Lu and Y), highlighting REE fractionation. Mollusks exhibited the highest REE concentrations, likely due to their habitat and feeding behavior, suggesting their potential as bioindicators of REE contamination.

Another assessment examined the transfer and fate of REEs in an impacted estuarine system in Brazil by analyzing sediments, suspended particulate matter (SPM), and organisms from different trophic levels (Santos et al., 2023). The highest ΣREE concentrations were found in sediments (180 ± 4.24 mg kg^{-1}) and SPM (163 ± 12.6 mg kg^{-1}). Phytoplankton (45.7 ± 5.31 mg kg^{-1}), periphyton (51.6 ± 1.81 mg kg^{-1}), and zooplankton (68.5 ± 1.27 mg kg^{-1}) were identified as key sources of REE exposure to the food web. However, REE concentrations were significantly lower in bivalves, crustaceans, and fish (6.01 ± 0.11, 1.22 ± 0.18, and 0.059 ± 0.003 mg kg^{-1}, respectively), with ΣREE decreasing with increasing trophic position, indicating trophic dilution. The findings suggest that seafood consumption is unlikely to be a major source of REE exposure for humans. However, due to multiple dietary sources of REEs, the authors suggest continuous monitoring to assess potential cumulative effects.

Another assessment examined REE distribution in different compartments of *Mytilus edulis* mussels from Norway, as well as in their ambient water and planktonic food (Zhang et al., 2025). Shale-normalized REY patterns in all compartments showed a decrease from light to heavy REY (LREE to HREE), without the REE depletion sometimes reported in the literature. Mussels accumulated REE at concentrations up to five orders of magnitude higher than in ambient water, preferentially taking up Ce and LREY over HREY. However, REE concentrations in mussels remained lower than those in plankton, with only minor fractionation except for the preferential uptake of La and Y. While metabolic REY fractionation within the mussels is minimal, shell formation is influenced by vital effects. Ce is rejected and decoupled due to its oxidation and the formation of Ce(IV) solution-complexes in the extrapallial fluid (EPF). HREE incorporation is limited by strong solution-complexation, whereas La, Gd, and Y are preferentially taken up due to their weaker complexes. The presence of siderophores in the EPF suggests that, unlike freshwater mussels (*Anodonta anatina*), *M. edulis* exhibits biological influences on REE anomalies. The authors indicate that this finding implies that these anomalies are partially biologically driven, limiting their reliability as paleo-redox proxies. Table 2.7 summarizes the main trophic web REE assessments and their findings.

Table 2.7 Summary of the main findings of studies concerning field-based trophic web REE assessments

Study	Location	Species/Organisms	REE Concentrations and Trends
Amyot et al. (2017)	14 Temperate Lakes, Canada	Fish, macroinvertebrates, zooplankton	ΣREE: Median bioaccumulation factors (log10) for fish muscle: 1.3, nonpredatory invertebrates: 4.4
Wang et al. (2019)	Maluan Bay, China	Seven common species (various taxa)	Light REEs (La to Eu) significantly more abundant than heavy REEs (Gd to Lu, Y)
Zhang et al. (2025)	Norway	*Mytilus edulis* (mussels), ambient water, planktonic food	Mussels accumulated REE up to five orders of magnitude higher than in ambient water, with Ce and LREY (La, Y) preferentially taken up
Santos et al. (2023)	Estuarine System, Brazil	Phytoplankton, periphyton, zooplankton, bivalves, crustaceans, fish	ΣREE in sediments: 180 ± 4.24 mg kg^{-1}; Phytoplankton: 45.7 ± 5.31 mg kg^{-1}; Zooplankton: 68.5 ± 1.27 mg kg^{-1}; Bivalves: 6.01 ± 0.11 mg kg^{-1}; Fish: 0.059 ± 0.003 mg kg^{-1}
Rétif et al. (2024b)	Seine Estuary, France	Primary producers (algae), mollusks, crustaceans, annelids, vertebrates	Total REE concentrations in primary producers and lower trophic levels: 0.14–156 µg g^{-1} dry weight; vertebrates: 0.03–0.15 µg g^{-1} dry weight

Chapter 3

Rubidium

Rubidium (Rb) is classified as a technology-critical element (TCE) due to its distinctive physical and chemical properties, including its high reactivity, excellent conductivity, and ease of ionization. These characteristics make it highly valuable in a range of advanced technological applications. As global demand for sophisticated technologies grows, rubidium's relevance has expanded significantly. It plays a key role in telecommunications, particularly in atomic clocks and global positioning systems (GPS), where its precision is critical. Additionally, Rb is used in electronics, specialty glass production, biomedical research, and quantum computing technologies. Its increasing importance is closely tied to the rising global reliance on high-performance materials, especially in fields requiring accuracy, stability, and miniaturization. However, as with many TCEs, concerns over supply security, limited natural availability, and the environmental impact of extraction and disposal are prompting calls for improved recycling strategies and sustainable resource management.

THE DISCOVERY OF RUBIDIUM

The metal Rb was discovered in 1861 by German chemists Robert Bunsen and Gustav Kirchhoff, in Heidelberg, Germany, while analyzing samples of the mineral lepidolite ($KLi_2Al(Al, Si)_3O_{10}(F, OH)_2$), a form of mica, through flame spectroscopy, a newly developed and groundbreaking method at the time (Kirchhoff and Bunsen, 1860). By analyzing the characteristic spectral lines produced when elements were heated to high temperatures in a flame, this method allowed for the precise detection and differentiation of even trace amounts of substances.

DOI: 10.1201/9781003412328-03

These chemists obtained Rb by extracting it from lepidolite through a series of chemical processes, including dissolution, precipitation, and electrolysis. First, they dissolved the lepidolite ore in acid and precipitated the potassium in the sample through distillation, which in turn carried down another heavier alkali metal. After washing the precipitate with boiling water the more soluble potassium (K) was removed and the atomic spectrum of the remaining component was noted as emitting two bright red lines in its emission spectrum, which had never been seen before, thus indicating a new element. Because of these bright red emission lines, Bunsen and Kirchhoff chose a name derived from the Latin word *rubidius* for this element, meaning "deepest red". It was then isolated in pure form by Bunsen in 1863 (Commission on Isotopic Abundances and Atomic Weights, 2015).

RUBIDIUM PROPERTIES

Rubidium belongs to Group 1 of the periodic table, the alkali metal group, with an atomic number of 37 and atomic weight of 85.47. It is a soft, silvery-white metal at room temperature, highly reactive and more so than its lighter alkali metal counterparts such as sodium and potassium, with an electronic configuration of [Kr] $5s^1$. It is the second least electronegative element, only behind francium, making it prone to forming ionic bonds and readily donating its valence electron in chemical reactions. Rubidium is among the most volatile alkali metals. Its low ionization energy (403 kJ mol^{-1}) contributes to its easy oxidation, making it highly reactive with water, where it forms rubidium hydroxide (RbOH) and hydrogen gas, often igniting spontaneously. The physicochemical properties of this element are depicted in Table 3.1 (RSC, 2025). Table 3.1 presents a comprehensive overview of the physicochemical properties of Rb.

Table 3.1 Physicochemical rubidium properties

Property	Description
Atomic number	37
Atomic weight	85.47
Density at 25°C	1.532 g cm^{-3}
Boiling point	688°C
Melting point	39.31°C
Bulk module	2.5 GPA
Poisson ratio	N/A
Thermal expansion coefficient from 20° to 100°C	N/A
Thermal conductivity	58 W mk^{-1}
Ionic radius	0.148 nm
Electronegativity	0.82

Roasting with limestone, at a mass ratio of 5:2 and temperatures between 800 and 1,000°C, resulted in poor extraction efficiencies for both lithium and Rb, likely due to insufficient $CaCO_3$. Better results were obtained with gypsum, which yielded 84% lithium extraction at 1,050°C, though Rb extraction remained low at under 14%. Sodium sulfate roasting showed the best lithium recovery, with approximately 90% extracted between 850 and 1,050°C, but Rb extraction was still unsatisfactory, peaking at 23%.

In another approach, sintering zinnwaldite with $CaCO_3$ (1:5 ratio) at 825°C followed by water leaching at 95°C achieved better extraction results: around 84% for lithium and 91% for Rb. During sintering, calcium carbonate-bearing minerals were formed, but the process was limited by the diffusion of lithium and Rb through a calcium hydroxide layer. Rubidium was later recovered as alum through crystallization, though this method produced a large amount of leach residue, which poses a challenge for process efficiency.

RUBIDIUM EXTRACTION FROM POLLUCITE

Pollucite, a colorless zeolite mineral with Rb and K as common substituents, is typically found in granite pegmatites and is often asso-ciated with minerals like lepidolite, spodumene, and albite. Major producers include Canada and Zimbabwe, with pollucite processing carried out in countries like the United States, Japan, Germany, and Russia.

Two common methods are used to decompose pollucite: acid digestion and roasting-water leaching, with acid digestion being more widely employed. In this method, pollucite is digested in hot hydrochloric acid, and Rb and cesium are recovered through precipitation with salts such as stannic chloride or iodine chloride. Alternatively, sulfuric acid can be used to crystallize Rb and cesium as alums, which can be decomposed by roasting and leaching to yield Rb and cesium sulfates.

The roasting-water leaching method involves roasting pollucite with substances like sodium carbonate or calcium chloride at high temperatures (800–900°C), followed by water leaching of the calcined material. The resulting Rb and cesium sulfates can then be converted into other salts via ion exchange or solvent extraction processes.

RUBIDIUM EXTRACTION FROM OTHER MINERALS

A variety of other Rb-bearing minerals have been investigated for Rb extraction using different processes.

Kaolin ore: Zheng et al. (2016) proposed a chlorination roasting process to recover Rb from kaolin ore. Using a mass ratio of kaolin to $CaCl_2$ of 2:1 at 800°C, a Rb extraction efficiency of 96.9% was achieved. Muscovite and kaolinite transformed into Ca ($Al_2Si_2O_8$) during roasting. Zhou et al. (2015) compared direct acid leaching and chlorination roasting-water leaching of kaolin clay waste. Sulfuric acid leaching with CaF_2 addition yielded an extraction efficiency of over 95%, while hydrochloric acid leaching achieved only 30%.

Gold waste: Tavakoli Mohammadi et al. (2015) studied Rb extraction from gold waste, which contains quartz and albite. Using a mass ratio of GW/Na_2SO_4/$CaCl_2{\cdot}2H_2O$ of 1:0.11:0.45 at 910°C, Rb extraction efficiency reached ~91%.

Muscovite ore: Shan et al. (2013) showed that a mixture of NaCl and $CaCl_2$ yielded a Rb extraction efficiency of 90.1% from muscovite ore through chlorination roasting at 850°C. Tian et al. (2020) achieved 96.7% Rb extraction via roasting at 800°C with a muscovite/$CaCl_2$ mass ratio of 2:1, followed by leaching at 60°C.

Biotite: Zeng et al. (2019) compared roasting-water leaching and direct acid leaching of Rb-bearing biotite. A Rb extraction of 97.2% was obtained at a biotite-to-$CaCl_2$ mass ratio of 2.5:1 at 900°C.

Rb-bearing granitic rock: Southern China hosts Rb-bearing granites that are complex to exploit. Xing et al. (2018) developed a sulfuric acid baking, reductive decomposition, and alkaline leaching process, achieving 94% Rb and 92% potassium recovery.

Polylithionite: Lv et al. (2020) reported extracting lithium and Rb from polylithionite via alkaline pressure leaching and solvent extraction. They achieved Rb and lithium recovery efficiencies over 95% at 250°C using NaOH.

Respirable sintering dust: Tang et al. (2018) recovered Rb from sintering dust through water leaching and solvent extraction. With a Rb content of 0.3%, Rb and potassium extraction efficiencies reached 95.2% and 93.4%, respectively, under optimal conditions. Rb was extracted using t-BAMBP and sulfonated kerosene.

RUBIDIUM EXTRACTION FROM BRINES

Salt Lake Brines

The current exploitation of salt lake resources focuses on extracting K, magnesium (Mg), and Li, but Rb and Cs are rarely recovered at an industrial scale, leading to resource wastage. Because of this, researchers have explored methods like solvent extraction and adsorption for Rb and cesium extraction from salt lake brines/simulated solutions.

Substituted phenols, such as t-BAMBP and BAMBP, are promising extractants for alkali metals, following the extraction order Cs > Rb > K > Na > Li (Arnold et al., 1965). Li et al. (2014) demonstrated that sodium did not affect Rb extraction, while potassium negatively influenced it, decreasing Rb extraction efficiency as its concentration increased. In another study by Liu et al. (2015), magnesium was precipitated before solvent extraction, while potassium and sodium were removed through scrubbing. This process achieved over 95% Rb and cesium extraction. Wang et al. (2015) optimized Rb extraction from lithium-bearing brine with t-BAMBP, using a pH of 13 and a 4:1 stoichiometric ratio of t-BAMBP to Rb. Li et al. (2017) achieved over 99% Rb and cesium extraction from simulated salt lake brine after five-stage countercurrent extraction, with a 19.4% co-extraction of potassium. Luo et al. (2019) selectively extracted Rb using an ionic liquid (IL) and bis(2-propyloxy) calix[4]crown-6, stripping Rb with NH_4Cl.

Adsorption is another technique used to extract Rb from simulated salt lake brine. Li et al. (2014) utilized a potassium titanium silicate calcium alginate composite adsorbent, achieving a Rb adsorption capacity of 1.55 mmol g^{-1}. However, the adsorption capacity decreased with higher NaCl concentrations. Functionalized metal-organic frameworks (MOFs) also show high adsorption capacities. For instance, Phenol@MIL-101(Cr) had a Rb adsorption capacity of 93 mg g^{-1} and could be regenerated with NH_4NO_3. Another MOF composite, sulfhydryl-Fe_3O_4@MIL-53(Al), exhibited a Rb capacity of 86 mg g^{-1} (Tian et al., 2019), with adsorption data following the Freundlich isotherm model. Dai et al. (2018) synthesized a phosphomolybdic acid-promoted MOF, achieving a Rb adsorption capacity of 99 mg g^{-1}. Ye et al. (2009) prepared a composite adsorbent using ammonium molybdophosphate, sodium alginate, and calcium chloride, extracting Rb and cesium through ion exchange reactions.

Sea Water Brines

Rubidium extraction from seawater brine is more challenging due to its lower Rb content. Researchers have used hexacyanoferrates of transition metals (KMFC, M = Cu, Ni, Co, Fe) for Rb extraction. Naidu et al. (2016a) investigated Rb adsorption using potassium cobalt hexacyanoferrate (KCoFC), which showed a Rb adsorption capacity of 96.2 mg g^{-1}, much higher than Li, Na, or Ca. Adsorption kinetics followed a pseudo-second-order model, with high desorption rates achieved using KCl. Naidu et al. (2016b) found that KCuFC exhibited superior Rb adsorption capacity compared to other hexacyanoferrates, likely due to more exchangeable potassium in the crystal lattice and higher surface potential for Rb adsorption. Naidu et al. (2018) reported that using polymer-encapsulated

KCuFC in a membrane-clinoptilolite zeolite adsorption system allowed for effective Rb recovery from simulated seawater reverse osmosis (SWRO) brine, achieving 83% Rb adsorption after potassium was reduced.

In another assessment, Choi et al. (2019) integrated membrane distillation with KCuFC adsorption to extract Rb from SWRO brine, with 0.45 mm KCuFC particles achieving the highest adsorption capacity. In addition to hexacyanoferrates, titanium silicate-based cation exchange resins ($M_2Ti_2O_3SiO_4{\cdot}nH_2O$) have been used for Rb extraction, with selectivity for alkali metals in the order Rb > Cs > K > Na > Li. Table 3.2 summarizes the main methods concerning Rb extraction from minerals and brines.

Table 3.2 Main methods concerning rubidium extraction from minerals and brines

Source	Extraction Method	Key Details
Lepidolite	Acid/alkali digestion, roasting-water leaching	Sulfuric or hydrochloric acid used for lepidolite dissolution, fractional crystallization for Rb and Cs.
Zinnwaldite	Roasting with limestone, gypsum, sodium sulfate; sintering with $CaCO_3$	Limestone yields poor Rb extraction; gypsum and sodium sulfate improve lithium recovery; sintering with $CaCO_3$ better for Rb.
Pollucite	Acid digestion (HCl, H_2SO_4); roasting-water leaching	Hydrochloric acid or sulfuric acid used for digestion; sodium carbonate or calcium chloride for roasting.
Kaolin Ore	Chlorination roasting	Uses $CaCl_2$ at 800°C to achieve Rb extraction.
Gold Waste	Roasting with Na_2SO_4/ $CaCl_2{\cdot}2H_2O$	Achieves 91% Rb recovery through roasting at 910°C.
Muscovite Ore	Chlorination roasting with NaCl/$CaCl_2$	High Rb extraction at 850°C.
Biotite	Roasting-water leaching, acid leaching	High Rb extraction with calcium chloride at 900°C.
Polylithionite	Alkaline pressure leaching and solvent extraction	NaOH used for high Rb and Li recovery.
Sintering Dust	Water leaching, solvent extraction	High Rb recovery from sintering dust using t-BAMBP and sulfonated kerosene.
Salt Lake Brines	Solvent extraction (t-BAMBP, BAMBP), adsorption (MOFs, composites)	High Rb recovery using substituted phenols and adsorbents like MOFs and potassium titanium silicate.
Seawater Brines	Hexacyanoferrates adsorption (KCuFC, KCoFC), membrane distillation, titanium silicate-based resins	High Rb adsorption from SWRO brine using KCuFC and titanium silicate-based resins.

RUBIDIUM APPLICATIONS

Rubidium's unique properties make it valuable for certain scientific and technological purposes applications, as depicted in Figure 3.1. These include uses in semiconductors, atomic clocks, specialty batteries, glass and ceramics, chemical synthesis, medical imaging equipment and general research.

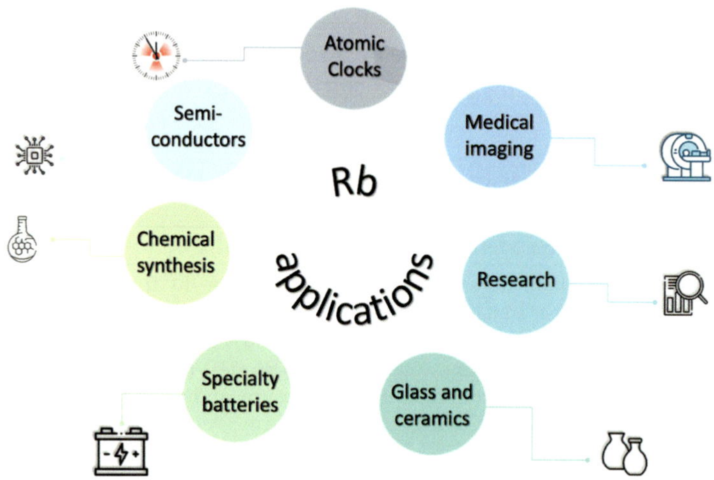

Figure 3.1 Main current rubidium applications.

The Rb isotope ^{87}Rb is commonly used in geochronology applications, employing radiometric dating techniques, such as Rb-strontium (Rb-Sr) dating, to determine the age of rocks and minerals. By measuring the ratio of ^{87}Rb to its decay product, ^{87}Sr, scientists can estimate the age of geological formations and events, as the decay rate is constant and well-known. Thus, by analyzing the ratio of these two isotopes, scientists can determine the time that has passed since the mineral or rock sample last cooled or crystallized. The process of dating Rb-bearing minerals via this technique began with the first age determination conducted by Hahn et al. in 1943. By the 1950s, the advent of mass spectrometers designed by Nier for isotope analysis of solids led to the widespread adoption of the Rb-Sr dating method (Bowen, 1994).

In addition, the bromine (Br) to Rb ratio (Br/Rb) has also been applied as a proxy for weathering processes in rocks, particularly to assess the degree of environmental chemical alteration and leaching, as both exhibit different behaviors during weathering, with Rb typically being more resistant to leaching compared to bromine. Thus, as rocks weather, Br tends to be more easily leached from minerals due to its solubility in water, whereas Rb is more likely to remain in the solid phase. Therefore, a higher Br/Rb ratio in weathered materials suggests a higher degree of chemical weathering, as more Br has been removed relative to Rb. Applying this ratio can aid in tracking the intensity and progress of rock weathering and provide insights into past environmental conditions, the origin of sediments, and the processes that have affected rock compositions over time (Wu et al., 2024).

Another ratio, between K and Rb, is an important tool for evaluating the extent of chemical weathering, with the inverse logic as for Br, where K forms stronger bonds with the minerals in rocks compared to Rb (Takashima et al., 2024). Because of this, Rb is more easily leached during weathering, whereas K tends to stay in the minerals longer. Thus, a higher K/Rb ratio suggests that more Rb has been leached out relative to K, indicating a greater degree of chemical weathering of the continental crust. This means that as weathering progresses, the amount of Rb decreases more quickly than potassium, leading to a higher K/Rb ratio. The increase in the K/Rb ratio generally corresponds with magmatic pulses (periods of volcanic activity), although there is a slight delay in this relationship. The trend in the K/Rb ratio also mirrors the changes observed in the angiosperm/gymnosperm index, which tracks the relative abundance of flowering plants (angiosperms) versus non-flowering plants (gymnosperms). This similarity suggests that both the K/Rb ratio and plant evolution patterns are influenced by similar geological or environmental factors over time.

The ^{87}Rb isotope is also used in atomic clocks, which utilize the precise oscillation frequency of electromagnetic radiation emitted by Rb atoms to maintain highly accurate time measurements (Bandi, 2024). These clocks rely on the hyperfine transition of rubidium-87 isotopes to maintain exceptional accuracy, making them smaller and more cost-effective alternatives to cesium atomic clocks. They are used in various scientific research, telecommunications, and navigation systems. They have, for example, been instrumental in advancing technologies such as the Global Positioning System (GPS), which relies on signals from satellites to pinpoint locations on Earth. The accuracy of these systems depends on precise time-keeping, and atomic clocks are crucial for synchronizing the clocks on GPS satellites with those on receivers on the ground. Atomic clocks also play significant roles in various industries. In telecommunications, for instance, they ensure the precise timing necessary for the effective transmission of data over long distances. In the cable TV sector, atomic clocks synchronize video signals to avoid timing discrepancies between video and audio. In the utilities industry, atomic clocks help regulate and monitor the power grid's frequency for the reliable operation of electrical devices, reducing the risk of power outages and ensuring stable electricity supply.

Furthermore, the ^{87}Rb isotope is also commonly used in experimental physics, particularly in studies involving atomic and molecular interactions, such as tests of the weak force and studies of parity violation (Altuntaş et al., 2018). Its radioactive decay properties also make it useful in experiments aimed at understanding the behavior of subatomic particles and interactions, *i.e.*, its role in Bose-Einstein condensation experiments has provided significant insights into quantum mechanics, such as the collective behavior of particles at extremely low temperatures.

Certain Rb compounds, such as rubidium fluoride (RbF), are also used in the manufacture of specialty glasses with specific optical properties (Shelby, 2021). These specialty glasses are highly valued in a variety of high-tech industries, especially in the fields of optics and photonics. Rubidium fluoride, due to its distinct refractive index and chemical stability, is used in the production of lenses, optical filters, and other optical components that require precise and high-quality performance. These materials are essential for devices such as lasers, which rely on specific light transmission characteristics, and for optical filters, which are used to selectively transmit light at particular wavelengths. The presence of Rb in these glasses enhances their ability to manipulate light with exceptional precision, which is critical in applications ranging from telecommunications to advanced medical instruments like endoscopes and microscopes.

In addition to their use in optical technologies, certain Rb compounds have aesthetic and entertainment applications, particularly in the pyrotechnic industry. In this sense, Rb compounds are sometimes used to generate distinctive and vibrant colors in fireworks. For instance, Rb chloride (RbCl) produces a striking purple flame when burned, produced by the excitation of Rb ions when exposed to heat, making it a popular choice in fireworks displays that aim to create visually impressive color schemes (Koch, 2002).

While not a direct medical application, certain Rb isotopes have been used in biomedical research for imaging studies and as tracers in certain physiological processes. For example, ^{82}Rb is used in positron emission tomography (PET) imaging for diagnosing heart conditions, administered intravenously to visualize blood flow to the heart and diagnose coronary artery disease. This application comprises a powerful tool for diagnosing obstructive coronary artery disease (CAD) because of its consistently high accuracy. In this sense, one study aimed to develop a normal database for quantifying myocardial perfusion and diagnosing obstructive coronary artery disease (CAD) using low-dose Rb-82 three-dimensional (3D) positron emission tomography (PET)-CT (Kaster et al., 2012). Out of a record of 1,501 patients, 77 were identified as having a low likelihood (LLK) of CAD. From this group, 40 LLK patients were used to construct a normal database with 4DM-PET, while the remaining patients were used to validate the normalcy of the database. In addition, 70 patients with confirmed CAD who had undergone both invasive coronary angiography and PET-CT were included to assess the database's accuracy in detecting CAD, using the sum-stress-score. The impact of clinical exclusion criteria and the inclusion of LLK patients were evaluated in this context. The results showed that the normal database for CAD detection had a 95% normalcy rate, and a 100% sensitivity for detecting patients with either 50% or 70% stenosis. Optimal specificity was achieved at 87% for both levels of stenosis. When localizing disease at 50% stenosis in the left anterior descending, left circumflex, and right coronary artery, sensitivity ranged from 59 to 68%, with specificity between 87% and 89%. At 70% stenosis, sensitivity improved to a range of 64 to 79%, while specificity ranged from 87 to 91%. In conclusion, this normal database, comprising relative perfusion scores of patients with LLK of CAD, can effectively diagnose obstructive coronary disease using low-dose Rb-82 with 3D PET-CT imaging.

Rubidium is also occasionally used in the production of semiconductors, particularly in the field of electronics. It can be employed as a dopant in certain semiconductor materials to modify their electrical properties. Rubidium vapor stabilization in semiconductor lasers involves using the specific absorption lines of Rb atoms to stabilize and control the frequency of the laser output (Furuta and Ohtsu, 1989). This technique exemplifies the well-defined and stable absorption spectra of Rb atoms to achieve high precision in laser frequency stabilization.

Rubidium isotopes are also used in various scientific studies as environmental tracers. In one of the first studies in this regard, Nirel and Reclavier (1999) carried out an assessment demonstrated the potential of utilizing Rb to Sr ratios for assessing pollution in the Allodon River basin, which is characterized by calcareous geology in France and Switzerland and has extremely low background dissolved Rb/Sr ratios. The findings indicate that this ratio represents an efficient diagnostic tool for MWW impacts in low Rb/Sr background waters, as the Rb/Sr ratio remains consistent regardless of river flow. Additionally, the authors also indicate that the Rb/Sr ratio can be utilized to model a watershed's ability to tolerate sewage effluent and to return to its natural quality.

In a similar study, the dissolved Rb to strontium ratio was applied as a conservative tracer for wastewater effluent-sourced contaminant inputs near a major urban wastewater treatment plant to substitute the determination of Gadolinium (Gd) as a tracer for contamination (Smith et al., 2021). Surface water samples from the tidal-fresh Potomac River near the Blue Plains Advanced Wastewater Treatment Plant (BPAWWTP) were collected along with treated MWW effluent samples. The results indicate that the Rb/Sr ratio was significantly enriched in MWW effluent

be the underlying reason for limited research interest in this element, as regulatory frameworks often drive scientific studies by prioritizing elements with established toxicological risks or economic significance. Without legal mandates or recognized environmental and health concerns, this element has likely been overlooked in favor of better-known contaminants, despite its potential ecological and toxicological implications in emerging contexts such as novel technologies and environmental contamination.

Some researchers have, in this context, highlighted the need for more focused studies on Rb in environmental assessments, acknowledging that the paucity of data on this element hinders a full understanding of its environmental impact. Beauvieux et al. (2024), for example, specifically noted the limited research on Rb in aquatic ecosystems and called for more comprehensive monitoring of its concentrations in freshwater and marine environments. Without targeted studies, it remains challenging to draw conclusions about Rb potential risks to aquatic life and human health, particularly in regions where Rb concentrations may be higher due to natural processes or anthropogenic activities, such as industrial use or mining.

In this regard, one study conducted an ecological risk assessment by determining the concentrations of several elements, including Rb, in recent sediments at Qaroun Lake, Egypt (Mahmoud et al., 2024). The study aimed to evaluate the potential environmental risks posed by various metals and metalloids in the lake, which has been impacted by agricultural runoff, industrial discharges, and other anthropogenic activities. Rubidium concentrations in the lake sediments ranged from 21 to 52 mg kg^{-1}, with notable variation observed between different zones of the lake, namely the eastern, middle, and western regions. The authors indicate that the detected spatial variability in Rb levels could be indicative of differing local environmental conditions, such as water chemistry, sediment characteristics, and proximity to potential pollution sources. While the study provides valuable information on the distribution of Rb in the sediments of Qaroun Lake, it does not discuss the specific sources or pathways through which Rb enters the lake ecosystem. The lack of discussion on the origins of Rb contamination is a significant gap also noted in other assessments, as understanding the source of this element is crucial for assessing the broader environmental and ecological risks associated with its presence.

In another sediment investigation, the extent of contamination and identify the sources of nine potentially toxic metals (PTMs) in alpine creek sediments near a uranium tailing dam in China were investigated, including Rb (Deng et al., 2024). Among the measured PTMs, Rb was found to be significantly enriched in the sediments, with an average concentration of 565.04 mg kg^{-1}. This value represents an exceedance of

100% compared to the background soil levels of Guangdong province, China. Such high Rb concentrations, along with other metals like Cd, Mn, Zn, and Sr, highlight the extent of contamination in the area. Using the Nemeiro integrated pollution index (NPI), Rb (with a pollution index value of 5.89) was identified as one of the metals posing a high environmental risk, alongside Mn and Cd. This index suggests that Rb contamination is of significant concern in the region, requiring focused attention for remediation efforts. Similarly, the Geo-accumulation index (Igeo) revealed that Rb presents a light pollution risk, with an Igeo value of 1.85, indicating its notable enrichment in the creek sediments compared to natural levels. The Enrichment Factor (EF) analysis further highlighted Rb as one of the most enriched metals, with an EF value of 16.75, surpassing the threshold of 10, which indicates significant anthropogenic influence. The enrichment of Rb, along with metals such as Cd and Li, is primarily attributed to the legacy of uranium hydrometallurgical activities in the region, which have led to the accumulation of these metals in the surrounding sediments. The findings reported by the authors highlight the substantial ecological risks posed by the high concentrations of Rb in the creek sediments.

In another assessment, the potassium: rubidium ratio of the sediments from the Canadian Arctic Archipelago was found to be higher than that of the source-area soils. This suggests that Rb behaves differently compared to potassium during the sedimentation process, which the authors postulate as likely due to selective ion-exchange reactions as the sediments transition from terrestrial to marine environments (MacDougall and Harris, 1969). Table 3.3 summarizes the main findings in studies concerning rubidium in sediments.

With regard to laboratory-based experiments, a two-compartment, clearance volume-based model was employed to investigate the pharmacokinetics of Rb and Cs in channel catfish (*Ictalurus punctatus*) following intravascular administration (Peters et al., 1999). The study compared the apparent volumes of distribution and clearances of both metals at 20.0°C and 27.5°C. The biological half-lives of Rb were determined as 15 to 16 days at both temperatures, while Cs had half-lives of 101 days at 20.0°C and 85 days at 27.5°C, five to seven times longer than those of Rb. This stark difference in half-lives suggests that Rb is more readily eliminated from the body of channel catfish compared to cesium, likely due to differences in the metabolism, uptake, and excretion mechanisms of the two metals. Both the intercompartmental and total body clearances of Rb were six to seven times greater than those of Cs. This indicates that Rb is distributed and eliminated more efficiently throughout the fish's body, which may be indicative of its lower affinity for storage in tissues compared to cesium. The central compartment volumes of distribution for Rb were twice those of Cs and remained constant with temperature changes. Peripheral compartment volumes for both elements were large compared to central compartments and decreased with higher temperatures. This suggests that while both Rb and cesium can accumulate in peripheral tissues, the effect of temperature on this accumulation is more pronounced in the case of cesium. Cesium was most concentrated in white muscle, with over 85% of this metal present there, and partitioning in peripheral tissues decreased with temperature. This preference for white muscle could be linked to the metal's interaction with muscle tissue-specific ion channels and transporters, which may preferentially accumulate cesium. However, the partitioning of cesium in peripheral tissues decreased as the temperature increased, indicating that temperature may play a role in altering the affinity of cesium for certain tissues. The authors concluded that these results demonstrate that pharmacokinetic modeling can reveal important aspects of radionuclide uptake and elimination.

Table 3.3 Main findings in studies concerning rubidium in sediments

Study Area	**Type of Sediment**	**Main Findings on Rb**	**Reference**
Qaroun Lake, Egypt	Recent sediments	Rb concentrations ranged from 21 to 52 mg·kg^{-1} across eastern, middle, and western zones. Spatial variability linked to local environmental factors. However, the study lacked discussion on Rb sources or pathways.	Mahmoud et al. (2024)
Alpine creek near uranium tailing dam, China	Creek sediments	Rb significantly enriched (mean = 565.04 mg·kg^{-1}), >100% above background. NPI = 5.89 (high risk), Igeo = 1.85 (light pollution), EF = 16.75 (significant anthropogenic input). Pollution linked to legacy uranium hydrometallurgy.	Deng et al. (2024)
Canadian Arctic Archipelago	Marine/ terrestrial sediments	K:Rb ratio in sediments higher than in source soils, suggesting Rb undergoes selective ion-exchange during transition from land to sea. Indicates differing geochemical behavior compared to K during sedimentation.	

Another study used Rb as a tracer to characterize potassium (K) uptake in *Carcinus maenas*, a species known for hyperosmoregulation in dilute seawater (Colburn et al., 2001). Hemolymph Rb levels were measured allowing for the determination of the initial rate of Rb uptake. The crabs were maintained for three weeks in dilute artificial seawater (35% ASW), where they successfully regulated their internal Na and K levels above those of the external media, showing elevated

Na-K-ATPase activity in the posterior gills.When Rb uptake was measured in assay conditions corresponding to 100% ASW, the rate of Rb uptake was similar in both 35% ASW crabs (0.45 ± 0.04 μmol g^{-1} h^{-1}) and crabs acclimated to normal seawater (100% ASW crabs, 0.49 ± 0.05 μmol g^{-1} h^{-1}). However, when measured under conditions that matched the 35% ASW environment, Rb uptake was significantly greater in 35% ASW crabs (0.28 ± 0.03 μmol g^{-1} h^{-1}) compared to 100% ASW crabs (0.10 ± 0.04 μmol g^{-1} h^{-1}). The difference in uptake rates between crabs from the two salinities was attributed to the combined effects of lower external Rb concentration and reduced salinity. The authors, thus, demonstrated that whole-body Rb uptake can be effectively measured using cation chromatography, and highlights that Rb uptake is greater in crabs living in low salinity (35% ASW) conditions compared to those in normal seawater.

A similar study investigated Rb as a tracer for potassium (K) to assess ion movement in seawater-adapted rainbow trout (*Salmo gairdneri*) and sculpin (*Leptocottus armatus*) (Sanders and Kirschner, 1983). The results showed that Rb and K behaved similarly in their exchanges between blood and tissues, as well as between blood and the external medium. More than 90% of an injected dose of either ion disappeared from the bloodstream within an hour, with full equilibration achieved after about 5 hours. During this time, the kinetics of blood concentration decrease were nearly identical for both ions. In trout, most tissue pools had specific activities of Rb (expressed as counts per minute of ^{86}Rb per μeq of tissue K) close to those in the blood plasma, with skeletal muscle as the exception, as its activity remained below 15% of plasma levels even several hours post-injection. When ouabain was administered or when external K concentrations were suddenly increased, both Rb and K were lost from cells at similar rates. The study also found that Rb and K were rapidly exchanged across the gills. Double-labelling experiments showed that the influx and efflux of K were about 1.27 times higher than those of Rb in trout, a value consistent with efflux data in sculpin. When fish were placed in seawater lacking Na and K, both ^{86}Rb and ^{42}K effluxes dropped by 75–80%, and gill membrane potentials shifted significantly—from positive to negative values. Reintroduction of either K or Rb led to partial repolarization, with K being slightly more effective. The authors conclude that the study demonstrated that ^{86}Rb is a suitable tracer for studying K transport in marine fish, although minor differences in membrane permeability require correction factors (e.g., multiplying Rb flux by 1.3) to accurately estimate K movement, especially across the gills. Table 3.4 summarizes the main findings on rubidium in laboratory-based assessments in aquatic organisms.

Table 3.4 Main findings on rubidium in laboratory-based assessments in aquatic organisms

Species/System	Matrix/System	Main Findings on Rb	Reference
Channel catfish (*Ictalurus punctatus*)	Intravascular pharmacokinetics	Rb has a biological half-life of 15–16 days (vs. 85–101 days for Cs). Rb showed higher total body and intercompartmental clearance, and larger central volume of distribution. Rb is more rapidly eliminated and less tissue-retentive than Cs. Peripheral compartment volumes decreased with temperature.	Peters et al. (1999)
Green crab (*Carcinus maenas*)	Hemolymph Rb uptake	Rb uptake used to trace K transport. Crabs in 35% ASW had significantly higher Rb uptake under low-salinity conditions. Demonstrated elevated Na^+/K^+-ATPase activity in posterior gills. Rb uptake reflects osmoregulatory adjustments in dilute environments.	Colburn et al. (2001)
Rainbow trout (*Salmo gairdneri*) and sculpin (*Leptocottus armatus*)	Blood, tissues, gills	Rb mimics K in kinetics. Over 90% of Rb left bloodstream within 1 hour; full equilibration in 5 hours. Rb and K showed nearly identical efflux/influx, including across gills. Ouabain and ion-free seawater reduced Rb/K efflux by 75–80%. Rb is a reliable tracer for K, but flux data needs a correction factor (~1.3×) for accuracy.	Sanders and Kirschner (1983)

Concerning field biota investigations, the occurrence and vertical distribution of several metals, including Rb, were investigated in the skull bones of Alpine bullheads (*Cottus poecilopus*) from the Javorinka river, located in the West Carpathians (Perinajová et al., 2020). Samples focused on the relationship between element concentrations and the fish's altitude and size. The authors report that calcium (Ca), Rb, and molybdenum (Mo) levels in the fish's skull were higher at higher altitudes and in smaller fish. In contrast, manganese (Mn), Sr, and zinc (Zn) concentrations increased with altitude. Rb, along with calcium (Ca) and molybdenum (Mo), was more concentrated in smaller fish found in higher, less accessible areas, while larger fish, found at lower altitudes, had lower levels of these elements. The authors indicate that this pattern suggests that as bullheads grow larger, their calcium, Rb, and molybdenum concentrations, supporting earlier findings that younger fish tend to have higher levels of molybdenum and Rb, as observed previously in rats (Kapitola and Vilimovska, 1984).

Another assessment by the same research group investigated the spatiotemporal distribution and transport of several metals, including Rb, from alpine terrestrial ecosystems to alpine lake and mountain stream populations of Alpine bullheads (Janiga et al., 2024). Rubidium levels in fish from alpine lakes and mountain streams were similar, suggesting comparable sedimentation patterns. Following a major flood in July 2018, Rb concentrations in fish dropped significantly in the year after the flood. However, Rb levels increased between 2020 and 2022 before declining again. The study also noted that Rb concentrations behaved similarly to Mo, showing a sharp decline post-flood and then a subsequent increase. The research found that Rb and other metals like strontium (Sr) remained in alpine lakes longer than macrominerals, influencing their accumulation in fish. The findings emphasize that Rb levels in fish are affected by environmental disturbances, such as flooding, and are influenced by the availability of other alkali metals in the environment.

In another study, Rb was one of the elements analyzed to assess its distribution among the muscle, liver, and kidneys in Indo-Pacific lionfish (*Pterois* spp.) collected from Cuban waters (Squadrone et al., 2020). Rubidium showed statistically significant differences in concentration among the three tissues. The highest Rb levels were detected in the muscle, followed by similar levels in the kidneys, with the liver displaying the lowest concentrations. This distribution pattern contrasted with that of most other trace elements, which generally accumulated more in the liver or kidneys. The elevated Rb levels in the muscle tissue highlight its relevance in dietary exposure assessments, particularly given the recent promotion of lionfish as a food source in the Americas and the Caribbean. The results demonstrated that while lionfish can serve as bioindicators for environmental contamination, especially through metal accumulation in liver and kidney, Rb's tissue-specific pattern provides insight into its distinct physiological handling and reinforces the need for tissue-targeted analyses in both ecotoxicological and food safety contexts.

In another assessment, Rb levels were determined in freshwater fish from Lake Erie and two Arctic lakes (Lake Hazen and Resolute Lake), as well as in seabirds, ringed seals, and invertebrates from the Northwater Polynya in Baffin Bay (Campbell et al., 2005). Additionally, two other alkali metals, cesium (Cs) and K, were also assessed in a subset of fish from Lakes Erie and Hazen. The reported findings indicate that Rb and Cs concentrations, along with Cs ratios in the sampled organisms, significantly correlate with $\delta^{15}N$ values, suggesting biomagnification through various food webs. In contrast, no significant relationship was observed between K concentrations and $\delta^{15}N$ values, indicating that K does not biomagnify and that homeostatic mechanisms regulate its levels. The authors conclude that Rb should, therefore, be considered alongside mercury and Cs in biomagnification assessments across different food webs and included in multi-element biomagnification assessments.

A similar assessment investigated the concentrations of several metals and metalloids, including Rb, along with stable carbon and nitrogen isotope analyses from various organisms in the Mekong Delta's main stream near Can Tho, South Vietnam (Ikemoto et al., 2008). A significant increase in Hg, Rb and selenium (Se) concentrations was observed with increasing $\delta^{15}N$ values, suggesting the biomagnification of these elements. The authors also reported that biomagnification patterns differed between crustaceans and fish, where crustaceans had higher levels of most determined elements, while fish presented high levels of Cr, Rb, and Hg. The authors suggest that variations in biomagnification profiles among major taxa may be due to differences in metal accumulation and detoxification mechanisms, such as the presence of metal-binding proteins like metallothionein. Metallothioneins (MTs) are low-molecular-weight, cysteine-rich proteins that have a high affinity for binding metals, including potentially Rb. These proteins play a critical role in the regulation of metal homeostasis within organisms, facilitating the sequestration and detoxification of excess metals, thereby minimizing their toxic effects on cells and tissues (Hogstrand and Haus, 1990; Ahearn et al., 2004). The ability of an organism to produce and regulate MTs can significantly influence its capacity to accumulate metals from the environment. For example, species with higher levels of MTs or more efficient metal-binding systems may be able to tolerate higher concentrations of metals in their tissues without suffering from the acute toxic effects that could otherwise occur in organisms with less effective detoxification mechanisms. Consequently, species with more robust MT-based detoxification processes may show lower levels of biomagnification, as metals are sequestered and detoxified before they can accumulate in significant amounts in tissues that are later consumed by higher trophic levels. Thus, further studies associating these metalloproteins to Rb concentrations in aquatic biota would be interesting to carry out aiming at furthering knowledge on this potential Rb-detoxification pathway.

Rubidium determinations have also been applied to differentiate between trophic levels in aquatic ecosystems. One study, for example, explored Rb/Cs ratios in juvenile sockeye salmon (*Oncorhynchus nerka*), threespine stickleback (*Gasterosteus aculeatus*), and both benthic and limnetic sticklebacks to differentiate between zooplankton feeders and benthic prey consumers (Chiasson, 1991). Only the benthic and limnetic sticklebacks exhibited Rb/Cs ratios influenced by their diet. Specifically, limnetic sticklebacks, which primarily feed on zooplankton, had higher Rb/Cs ratios compared to benthic sticklebacks that consume prey from the bottom, such as invertebrates and small fish. This difference in the ratios is likely due to the distinct metal signatures associated with the organisms from each habitat, as zooplankton and benthic prey species might accumulate Rb and Cs in different proportions due to variations in their environments and metabolic processes. The Rb/Cs ratio thus serves as a biomarker for feeding habits, providing a tool for ecologists to infer the dietary sources of different species without needing to directly observe their feeding behavior. Interestingly, the study also found that the Rb/Cs ratios in juvenile sockeye salmon were influenced by the concentrations of these elements in the egg yolk during early development, rather than directly reflecting the fish's diet. This highlights the complex nature of metal accumulation in organisms, where early-life exposure can have lasting effects on the elemental composition of tissues. The elemental profile of the yolk, which is rich in nutrients and metals absorbed from the maternal environment, may therefore set the baseline for the fish's subsequent accumulation of Rb and Cs as they mature. For threespine sticklebacks residing in both littoral and limnetic zones of the same lake, the data indicated that the fish needed to inhabit each zone for several months before their Rb/Cs ratios in tissues accurately reflected the concentrations found in their prey. This temporal delay emphasizes the dynamic nature of metal accumulation and the importance of considering habitat residency and time spent in different environments when interpreting Rb/Cs ratios. In these fish, the accumulation of Rb and Cs in tissues is influenced not only by the immediate diet but also by the broader environmental context, as these elements are absorbed over time through both food and water exposure.

Interestingly, several Rb assessments have been recently conducted in top predators or mesopredators, such as sharks. This is important to obtain potential insights into the movement of Rb through food webs and its bioaccumulation across trophic levels. By assessing Rb concentrations in this group, a range of ecological dynamics can be investigated to deepen understanding on ecosystem function, pollutant cycling, and aquatic environment health.

In the first assessment carried out in a coastal shark, Rb levels, alongside other elements, were determined in Atlantic nurse sharks (*Ginglymostoma cirratum*) from the state of Maranhão, Brazil (Wosnick et al., 2021). The study also evaluated selenium (Se) due to its potential protective effects and examined two non-enzymatic biomarkers: metallothionein (MT) and reduced glutathione (GSH). Findings revealed lower concentrations

Rb, and Se during the rainy season, no significant correlations Se and Rb, and no correlation MT or GSH and the determined metals, suggesting no MT-pathway detoxification and no induced antioxidant defenses. These findings indicate that Rb does not appear to influence or be influenced by Se levels in this species under the environmental conditions assessed. This absence of correlation may imply that Rb contamination does not interfere with the protective pathways mediated by Se or that Se is not being mobilized to mitigate potential Rb toxicity. Alternatively, it could reflect a lack of co-exposure to these elements at concentrations sufficient to elicit measurable interactions. The seasonal decrease in Se concentrations, coinciding with lower Rb levels, could suggest that both elements are influenced by environmental factors such as water dilution during the rainy season, rather than by biological mechanisms or synergistic interactions. Furthermore, Rb exposure does not seem to induce MT synthesis or that MT is not the primary pathway for Rb detoxification in these organisms. Unlike other metals that form strong bonds with thiol groups, Rb, being an alkali metal, may not interact directly with MTs, explaining the lack of association. Alternatively, the levels of Rb in the environment may have been insufficient to trigger MT-mediated detoxification processes. Finally, Reduced glutathione (GSH) is a key antioxidant molecule that protects cells from oxidative stress by neutralizing free radicals and repairing oxidative damage. The study found no significant correlation between Rb levels and GSH concentrations, indicating that Rb exposure did not induce oxidative stress in the nurse sharks under the conditions studied. This lack of response may be due to the relatively low concentrations of Rb detected, which might not have been high enough to generate ROS or disrupt redox homeostasis. Furthermore, the absence of a GSH response suggests that Rb contamination in this context is unlikely to pose immediate oxidative challenges to the species. However, it is also possible that other antioxidant systems, such as enzymatic pathways involving superoxide dismutase (SOD) or catalase (CAT), may play a more prominent role in mitigating any oxidative effects of Rb in this species. Thus, further assessments concerning Rb interactions with cellular defenses are required. The potential Rb contamination was linked to a local space rocket explosion, which, along with subsequent dredging activities, seems to have impacted environmental quality.

The second study regarding a coastal shark comprised a baseline report on Rb contamination in juvenile Caribbean Sharpnose sharks (*Rhizoprionodon porosus*) from southeastern Brazil (Hauser-Davis et al., 2023). An analysis of 18 juvenile sharks demonstrated that muscle tissue contained significantly higher Rb concentrations compared to liver tissue, suggesting a bioaccumulation process. Significant differences were also noted between sexes, with females presenting higher Rb concentrations in the liver. A negative correlation between total length and liver Rb, and a positive correlation between total length and muscle Rb in males, were observed, highlighting potential ecological concerns related to Rb contamination, which may affect growth and other traits in this species.

Another top-predator assessment encompasses the first report on the assimilation and accumulation of TCEs, carried out in tiger sharks (*Galeocerdo cuvier*) (Wosnick et al., 2024). The analyzed tiger shark samples from south Brazil presented significant Rb concentration variations across tissues, with higher levels detected in kidneys and lower levels in liver. Statistically significant correlations between Rb and several Rare Earth Elements were noted, suggest that these elements are systematically transported across different tissues.

Interestingly, Rb has also been reported as a constituent of shark teeth. One study reported Rb concentrations in bull shark (*Carcharhinus leucas*) teeth from Fiji (0.073–0.105 mg kg^{-1}; mean: 0.086 ± 0.007 mg kg^{-1}, without distinguishing enamel and dentin), Australia (enamel: 0.01 ± 0.00 mg kg^{-1}; dentin: 0.088 ± 0.01 mg kg^{-1}), and Lake Nicaragua (enamel: mean 0.03 mg kg^{-1}; dentin: mean 0.09 mg kg^{-1}), similar to those observed in tiger sharks (Wosnick et al., 2024). However, bull sharks from Tonga exhibited lower Rb concentrations (enamel: mean 0.018 mg kg^{-1}; dentin: mean 0.0227 mg kg^{-1}) (Kocsis et al., 2015). In contrast, bull sharks from South Africa had higher Rb levels in enamel (0.02 ± 0.01 mg kg^{-1}) and dentin (0.34 ± 0.11 mg kg^{-1}), as did bull sharks from Florida (enamel: 0.041 mg kg^{-1}; dentin: 0.635 mg kg^{-1}) (Kocsis et al., 2015). The observed differences in Rb concentrations in shark teeth across locations may be due to environmental Rb availability, both due to local geological characteristics, influencing Rb content in local water bodies and human activities such as mining, agriculture, and industrial runoff, which can increase Rb levels in certain habitats. Dietary variations are also a factor, as sharks, as most animals, acquire Rb primarily through their prey. Furthermore, coastal regions often show more elemental variability due to terrestrial runoff compared to the more stable conditions of the open ocean. Physiological factors, such as tissue-specific deposition of Rb in enamel and dentin, growth rates, and life stage, may also influence Rb accumulation, as well as migratory or resident behaviors, as migratory species may integrate exposure from multiple regions, diluting extreme regional variations, while resident sharks reflect local environmental conditions. Teeth are, however, not well assessed in ecotoxicological assessments, and further evaluations concerning different species, the specific part of the tooth analyzed (e.g., enamel versus dentin), as well as including temporal and seasonal variations should be carried out to further understand the dynamics of this element in top-predator teeth. Table 3.5 summarizes the main findings of the aforementioned assessments.

Table 3.5 Main findings on rubidium biota assessments

Reference	Species/Location	Analyzed Tissue	Key Findings	Observations
Perinajová et al. (2020)	Alpine bullhead (*Cottus poecilopus*)/Javorinka River (Carpathians)	Skull bones	Higher Rb in smaller fish and higher altitudes	Similar patterns with Ca and Mo; suggests ontogenetic and altitudinal effects
Janiga et al. (2024)	Alpine bullhead/Alpine lakes and streams	Whole fish	Rb levels dropped post-flood, later rose	Parallels with Mo; Rb retention influenced by sedimentation and flooding
Squadrone et al. (2020)	Indo-Pacific lionfish/Cuba	Muscle, liver, kidneys	Highest Rb in muscle	Important for food safety; differs from common trace metal patterns
Campbell et al. (2005)	Fish, seabirds, seals/Arctic (Lakes Erie, Hazen, Resolute, Northwater Polynya)	Whole organisms	Rb correlates with δ15N (biomagnification)	K does not biomagnify; Rb to be included in multi-element food web studies
Ikemoto et al. (2008)	Fish and crustaceans/Mekong Delta	Whole organisms	Rb increases with δ15N	Biomagnifies; potentially influenced by metallothionein-based detoxification
Chiasson (1991)	Salmon, sticklebacks/Canada	Whole fish	Rb/Cs ratios reflect feeding habits	Limnetic > benthic sticklebacks; influenced by egg yolk and habitat
Wosnick et al. (2021)	Atlantic nurse shark/Maranhão, Brazil	Muscle, liver	Lower Rb in rainy season; no MT or GSH correlation	Suggests environmental dilution; no Rb detox via MT/GSH
Hauser-Davis et al. (2023)	Caribbean sharpnose shark/SE Brazil	Muscle, liver	Muscle > liver; sex and size differences	Suggests bioaccumulation and ecological relevance
Wosnick et al. (2024)	Tiger shark/South Brazil	Kidneys, liver, other	Kidneys > liver; Rb correlated with REEs	Suggests systemic transport of Rb and REEs
Various	Bull shark/Fiji	Teeth	0.073–0.105 mg kg^{-1} Rb in teeth	Indicates Rb incorporation into calcified structures

BIOLOGICAL RUBIDIUM EFFECTS

Rubidium has traditionally been considered a non-toxic element, with its relatively low environmental and biological reactivity often leading to its classification as benign in natural ecosystems. Its relatively mild toxicity profile has contributed to its limited inclusion in environmental monitoring and regulatory frameworks. However, as new technologies and industrial applications involving Rb continue to emerge, there is growing concern about the potential for overexposure to Rb compounds and the complex interactions it may have with other metals and pollutants, both in environmental and human health contexts.

At least one study has suggested Rb essentiality in animals (Yokoi, 1997), when investigating the effects of low dietary tin (Sn) and Rb levels on the growth, tissue mineral concentrations, and selected plasma biochemical indices in growing rats. Rats fed a diet containing 17 ng of Sn per gram showed reduced weight gain, decreased food efficiency, alopecia, altered tissue mineral levels, and changes in plasma lipid concentrations, compared to those given 2 μg of Sn per gram of diet. Similarly, rats on a 0.5 μg Rb per gram diet presented reduced Rb concentrations in all tissues, altered mineral levels in various tissues, and elevated plasma urea nitrogen levels compared to those on an 8 μg Rb per gram diet. These findings suggest that tin and Rb may be essential for animals.

Concerning effects in wildlife, one study observed that Rb may inhibit spermatogenesis and cause spermatogonia necrosis in fish (Yamaguchi et al., 2007). The authors sampled cultivated male catfish (*Pangasianodon hypophthalmus*) and examined testicular development and reported negative correlations between gonadosomatic index (GSI) and Rb concentrations. In addition, *in vitro* testicular organ cultures using Japanese eel (*Anguilla japonica*) were also conducted. While individual treatments with Rb did not affect spermatogenesis, exposure to 10^{-7} mol L^{-1} Pb associated to 10^{-5} to 10^{-3} mol L^{-1} Rb inhibited spermatogenesis induced by 11-ketotestosterone. The reported findings, thus, suggest that Rb has an inhibitory effect on spermatogenesis in catfish from the Mekong Delta.

Another study evaluated several elements, including Rb, in 135 juvenile gilthead seabream specimens, a commercially important species in the Mediterranean Sea for both aquaculture and fisheries, from seven coastal lagoons (Beauvieux et al., 2024). The results showed the accumulation of multiple trace elements in the seabreams, with distinct contamination signatures across lagoons. The potential adverse effects of the complex contamination mixtures on the liver (the main organ responsible for xenobiotic metabolism) and red muscle (a highly metabolic organ) using a proteomic approach were also assessed. Concerning Rb, this element seems to be involved in a downregulation of 52 enriched pathways, primarily associated with cytoskeleton and cellular organization.

In yet another assessment, Rb concentrations were determined in the octopus species *Octopus vulgaris* collected from three locations in Portugal (Cascais, Santa Luzia, and Viana do Castelo) during autumn and spring (Seixas and Pierce, 2005). The concentrations of Rb, along with vanadium (V) and potassium (K), were analyzed in various tissues, including the digestive gland, branchial heart, gills, mantle, and arms, for both males and females. The study found that Rb concentrations were present across the different tissues but did not show any significant differences between sexes. A significant positive correlation between Rb and potassium concentrations was observed, particularly in the branchial heart. Although vanadium was detectable only in the digestive gland and branchial heart, an important finding was the significant relationship between the concentrations of vanadium, Rb, and K in the branchial heart. The authors indicate that these results suggest that the branchial heart plays a crucial role in the accumulation and possible detoxification of these elements, including Rb, which may have implications for understanding the physiological processes of *Octopus vulgaris* in response to environmental contamination.

It is important to note that the interaction of Rb with other metals, such as those commonly found in contaminated environments, could lead to synergistic or antagonistic effects that might exacerbate the toxicological risks of exposure. This may, in turn, create complex exposure pathways that are not yet fully understood. Moreover, Rb could influence the bioavailability and uptake of other metals or be affected by changes in water chemistry and environmental conditions, further enhancing its toxicity in certain ecological settings. However, studies exploring these interactions, particularly the combined effects of Rb with other metals or environmental contaminants, are seemingly nonexistent, highlighting a significant gap in the understanding of its full ecological and toxicological risks.

Concerning Rb investigations in mammal models, one study investigated the effects of low dietary Rb on plasma biochemical parameters and tissue mineral levels in rats (Yokoi et al., 1996). Eighteen male Wistar rats, each weighing, were divided into two groups and given diets with or without added Rb (0.54 mg kg^{-1} versus 8.12 mg kg^{-1}) for 11 weeks. Rats on the Rb-supplemented diet exhibited lower plasma urea nitrogen levels, higher tissue Rb concentrations, and altered mineral levels compared to those on the diet without Rb. Specifically, the non-supplemented rats presented elevated urea nitrogen in plasma; reduced Rb in tissues; decreased sodium in muscle; increased potassium in plasma, kidney, and tibia, but lower potassium in testis; lower phosphorus in heart and spleen; reduced calcium in spleen; increased magnesium in muscle and tibia; higher iron in muscle; decreased zinc in plasma and testis; and lower copper in heart, liver, and spleen, with increased copper in kidney. The authors, thus, concluded that tissue Rb levels correlate with dietary intake and that Rb deficiency can disrupt the balance of several essential minerals including sodium, potassium, phosphorus, calcium, magnesium, iron, zinc, and copper.

In another study, various Rb compounds with different counter anions—acetate, bromide, carbonate, chloride, and fluoride—were orally administered as a single dose to several groups of rats (Usuda et al., 2014). Cumulative 24-hour urine samples were collected to measure Rb, fluoride, N-acetyl-β-D-glucosaminidase (NAG), and creatinine levels, assessing potential acute renal effects. Additionally, daily serum samples were analyzed for aspartate and alanine aminotransferase (AST and ALT) levels to evaluate possible acute hepatic effects. Urinary Rb excretion recovery rates of 8.0–10.5% suggest that urine can serve as a useful diagnostic tool for Rb exposure. The findings indicate that Rb's biological effects vary depending on the counter anion. Notably, rats treated with anhydrous Rb fluoride exhibited significant NAG leakage and elevated ALT levels, suggesting renal and hepatic toxicities, likely due to fluoride.

In an interesting approach, one study investigated the interaction of Rb with microbial commensals, in mice treated with Rb chloride (RbCl) (Chen et al., 2021). Sixty-four mice were divided into four groups: a blank control group receiving normal saline, and three experimental groups with low (20 mg L^{-1}), middle (50 mg L^{-1}), and high (100 mg L^{-1}) doses of RbCl. The mice were administered RbCl intragastrically twice daily for six weeks. Fecal samples were collected after the treatment period, and mice were sacrificed for organ weight analysis. Fecal DNA was extracted and sequenced to analyze gut microbiome changes using specific primers targeting the V4 region of the 16S rRNA gene. The sequencing data were processed to identify operational taxonomic units (OTUs) and assess microbiome diversity and richness. Results from 16S ribosomal RNA gene sequencing of fecal microbiota indicated that RbCl generally preserved the diversity of the fecal microbial community, though notable shifts in microbial composition occurred after RbCl exposure. Specifically, RbCl significantly increased the abundances of Rikenellaceae, Alistipes, Clostridium XlVa, and sulfate-reducing bacteria such as Deltaproteobacteria, Desulfovibrionales, Desulfovibrionaceae, and Desulfovibrio. Conversely, it notably decreased the abundances of Tenericutes, Mollicutes, Anaeroplasmatales, Anaeroplasmataceae, and Anaeroplasma lineages. Regarding archaea, only two less abundant genera, Sulfolobus and Acidiplasma, were observed. The authors suggest that changes in fecal microbiota could partially contribute to the anticancer or antidepressant effects of RbCl, as these clinical conditions are directly associated to fecal microbiota alterations, highlighting the potential of targeting the microbiome for therapeutic purposes.

Rubidium is currently not regarded as essential for biological functions in humans, and its biological role remains largely unclear (Usuda et al., 2014). Although trace amounts of Rb are naturally found in the human body, totaling about 350 mg, no specific affinity to any organ has been noted, although higher levels have been noted in the liver and muscle and lower levels in the skeletal system (Nielsen, 1996; Hass, 2006; Yokoi, 1997). Some studies have hinted at possible roles for this element in certain biochemical pathways, but further research is required to determine its importance and effects.

In humans, Rb has been correlated with phenylketonuria and maple-syrup-urine disease. In this sense, a study of 46 healthy children, Rb levels showed no significant differences between infants, toddlers, and school-aged children (Lombeck et al., 1980). In contrast, 29 patients undergoing dietary treatment for phenylketonuria and maple syrup urine disease exhibited significantly lower Rb levels compared to healthy children. During the initial three months of dietary therapy, Rb levels remained at the lower end of the normal range but subsequently dropped to approximately 60% of the average normal values. Over time, these levels continued to decline with prolonged dietary treatment.

Since Rb and potassium (K) are both alkali metals and have similar chemical properties, Rb can be taken up by cells in place of K (Williams and Leggett, 1987), mainly through K_{ATP} channels. This can potentially interfere with various cellular processes that rely on K ions. Studies in this sense, however, are still in their infancy, even though this fact has been known since the 1980s. In one of the first assessments concerning Rb similarities to potassium (K), K-channel openers in isolated aortas from rats were exposed to Rb *in vitro* (Greenwood and Weston, 1993). Potassium was replaced by Rb in a physiological salt solution (PSS) (RbPSS vs normal PSS) to study its effects on vascular smooth muscle, specifically rat aortic segments. Millimolar concentrations of Rb appeared to inhibit KATP channels. These channels are crucial for the action of K-channel openers, which cause relaxation of vascular smooth muscle by opening K_{ATP} channels, leading to hyperpolarization and reduced contraction. When Rb was present, the relaxation induced by K-channel openers was less potent, suggesting that this metal interferes with the opening of these channels. Rb did not lead to a detectable increase in the efflux of either Rb or K from the aortic tissues, unlike in normal PSS where significant ion exchange occurred. This suggests that Rb blocks the normal ion transport processes mediated by KATP channels. Rb also caused depolarization of the aortic tissue, which made the K-channel openers ineffective in changing the membrane potential. In normal conditions with K, these drugs would hyperpolarize the membrane, leading to muscle relaxation, but Rb prevented this effect. Thus, the authors reported that Rb seems to mimic K in some ways but also have distinct effects that inhibit normal K-channel function.

One recent study in this regard reported an Rb^+ efflux assay to evaluate K_{ATP} channel function and activity (ElSheikh et al, 2024). These channels are ATP-sensitive potassium channels, a type of ion channel found in the membranes of various cells, important for regulating the flow of potassium ions (K^+) across the cell membrane, and their activity is sensitive to the concentration of ATP (adenosine triphosphate) inside cells. Thus, this type of assay may present potential therapeutic applications. In another study, the gating and conductance properties of voltage-gated potassium channels formed by KCNQ1 alone (homomeric) or in combination with the small KCNE1 subunit (heteromeric) was conducted using *Xenopus* oocytes, an *in vitro* model (Pusch et al., 2000). These channels help control the heart's electrical activity, and alterations can cause health conditions, like long QT syndrome. Two-electrode voltage-clamp and patch-clamp techniques were used, along with solutions containing high concentrations of K or Rb, to explore how both ions affect channel behavior. For homomeric KCNQ1 channels, substituting K with Rb increased inward tail currents approximately threefold, despite Rb lower permeability, suggesting that Rb influences gating. However, changes in tail current kinetics and steady-state activation were minimal. Single-channel amplitude at negative voltages, estimated through nonstationary noise analysis, showed only a modest increase (1.2-fold) in Rb. Filtering the data at lower cutoff frequencies revealed a rapid flickery/block process that reduced apparent conductance, while the relative conductance in Rb compared to potassium increased at lower frequencies (about twofold at 10 Hz), likely due to reduced channel blockage. For heteromeric KCNQ1/KCNE1 channels, macroscopic inward tail currents in Rb were about two-fold smaller and exhibited a pronounced sigmoidal time course similar to the inactivation process seen in Rb KCNQ1 channels, suggesting multiple open states. The single-channel amplitude in Rb was roughly half that in potassium at a 5-kHz bandwidth, with lower cutoff frequencies reducing apparent conductance. However, the Rb-to-K conductance ratio remained constant across different filtering conditions. The findings indicate a rapid flicker process largely independent of gating. In homomeric channels, Rb occupancy at negative voltages stabilized the flicker-open state and slowed flickering, while in heteromeric channels, Rb did not alter flickering behavior. The effects of KCNE1 on conduction properties are consistent with its interaction in the channel's outer vestibule. These differences aid in understanding how KCNE1 changes the way the K channels work and might provide data on treating heart conditions linked to these channels.

It is interesting to note that certain human diseases, such as Alzheimer's disease, are marked by a loss of intracellular K compartmentalization, reflecting membrane damage and mitochondrial dysfunction. In this sense, one assessment analyzed K and Rb levels in the blood, cerebrospinal fluid (CSF), and brain tissue of Alzheimer's patients compared to healthy controls. A significant decrease in K and Rb was found in all intracellular compartments of the brain. However, Rb levels in red blood cells (erythrocytes) and CSF showed no significant differences between groups, with only a slight decrease in Alzheimer's patients compared to controls. The widespread reduction in cortical K and Rb suggests a potential energy crisis within the brain (Roberts et al., 2016).

In another human health assessment, several elements, *i.e.,* mercury (Hg), iron, Rb, selenium, and zinc levels were measured in the pituitary gland to explore differences in environmental Hg exposure between Alzheimer's patients and healthy controls (Cornett et al., 1998). The pituitary gland was chosen as it is considered a reliable indicator of mercury exposure. No significant differences were found in the levels of these elements between the groups, indicating that Alzheimer's patients did not have higher Hg exposure than controls.

Rubidium has been suggested as a potential anticancer agent, yet no studies have explored the relationship between Rb levels in biological samples and breast cancer risk or its potential as a biomarker for risk assessment. In one study, survey data and urine specimens were collected from 240 women with newly diagnosed invasive breast cancer, prior to treatment, and from 246 age-matched female controls (Su et al., 2011). Urinary Rb concentrations were measured using inductively coupled plasma mass spectrometry. Creatinine-adjusted Rb levels, expressed as median values with 25th and 75th percentiles, were significantly lower in breast cancer cases compared to controls. After adjusting for potential breast cancer risk factors, women in the second and highest tertiles of Rb concentration had a decreased risk of breast cancer in a dose-dependent manner compared to those in the lowest tertile. Rb concentrations were significantly and inversely associated with breast cancer risk and may serve as a potential biomarker for risk assessment. Additionally, women with higher urinary Rb levels showed a markedly decreased risk of developing breast cancer, highlighting the potential clinical significance of Rb in cancer prevention.

Since dialysis has extended long-term survival for uremic patients, more subtle issues, such as imbalances or suboptimal regulation of trace elements, have become of concern. In this sense, one study focused on Rb levels in dialysis patients due to concerns about potential deficiencies (Canavese et al., 2001). Rb status was evaluated in uremic patients through three methods: (A) Serum concentrations were measured using graphite furnace atomic absorption spectroscopy from blood samples of 70 chronic hemodialysis (HD) patients compared to 75 controls. (B) Tissue concentrations were assessed using neutron activation analysis from 20 autopsy or biopsy samples of HD patients compared to 21 controls. (C) In vivo intradialytic mass balance was measured during standard bicarbonate dialysis in eight HD patients. Results showed that serum Rb concentrations were significantly lower in HD patients compared to controls. Univariate logistic regression revealed a higher risk of serum Rb levels below 250–350 μg L^{-1} in uremic patients. Furthermore, tissue Rb levels were also significantly lower in HD patients, including in the brain. Finally, during a standard bicarbonate dialysis session, Rb was transferred from the blood to the dialysis bath, with a mean intradialytic Rb removal average of 4.0 mg/session. The reported findings confirm that Rb deficiency may occur in uremic patients, and that standard bicarbonate dialysis contributes to Rb removal. However, the removal rate is not significantly higher than what would be expected with normal urinary output (20 mg/week). Further research is needed to clarify the causes of Rb deficiency, including the effects of uremia, pre-dialysis factors (diet, impaired renal function, medications), and dialysis parameters (frequency, duration, and techniques). As Rb plays a role in neurobehavioral functions, particularly as an antidepressant, its deficiency may impact central nervous system function and quality of life in uremic patients. Monitoring serum Rb and identifying the mechanisms behind its deficiency could inform future therapeutic strategies.

Chapter **4**

Platinum and Platinum-Group Metals

Platinum and platinum-group metals (PGMs) include platinum (Pt), palladium (Pd), rhodium (Rh), ruthenium (Ru), iridium (Ir), and osmium (Os). These metals are collectively categorized as "platinum-like" due to a unique combination of shared physical, chemical, and structural characteristics that align closely with those of platinum. Their similarities stem largely from their position in the periodic table as transition metals occupying the d-block, specifically groups 8 to 10 and periods 5 and 6. As such, they exhibit comparable electron configurations, which influence their valency and bonding behavior.

Platinum and platinum-group metals are known for their high melting points, excellent corrosion resistance, and catalytic activity, properties largely attributable to their dense electron clouds and strong metallic bonding. They are chemically inert under many environmental conditions and display exceptional stability in both acidic and oxidative environments. These features make them indispensable in various high-performance applications, particularly in the automotive industry for catalytic converters, where they facilitate the reduction of harmful exhaust gases. Their catalytic properties are also widely harnessed in chemical synthesis, petroleum refining, and emerging green technologies such as hydrogen fuel cells and water electrolysis.

Structurally, PGMs form similar types of crystal lattices—primarily face-centered cubic (FCC) for Pt, Pd, and Ir—which contribute to their mechanical strength and resistance to deformation. Additionally, due to their shared geochemical behavior and frequent co-occurrence in natural ore deposits, PGMs are typically mined and refined together, further

DOI: 10.1201/9781003412328-04

reinforcing their collective classification. Despite their commonalities, each PGM has unique traits that suit specific technological demands, yet all remain critically important due to their scarcity, economic value, and key roles in advancing sustainable and high-tech innovations.

One defining feature of platinum and PGMs is their outstanding resistance to chemical corrosion, even at high temperatures, due to their low reactivity and high activation energy for oxidation (Seymour and O'Farrelly, 2012). Additionally, they share a remarkable catalytic efficiency, which stems from their partially filled d-orbitals that enable them to readily form complexes, activate molecular bonds, and participate in redox reactions. This makes them invaluable in applications such as automotive catalytic converters, hydrogen production, and industrial chemical processes.

Platinum-group metals also exhibit high melting points, densities, and electrical conductivities, which are typical of platinum. These properties result from the strong metallic bonds formed by their d-electrons and relativistic effects, which are particularly pronounced in the heavier elements like platinum, iridium, and osmium (Davies and Hartley, 1981). The combination of these properties gives them a unique versatility that is crucial for advanced technological and industrial applications.

The term "platinum-like" also reflects their geochemical association, as these metals are often found together in nature, either in sulfide ores or as native alloys associated with mafic and ultramafic rocks and their derivatives (O'Driscoll and González-Jiménez, 2016), due to their similar chemical behaviors. This natural co-location leads to their simultaneous extraction during mining and refining operations, typically through complex metallurgical processes designed to separate and purify each element. The shared geochemical pathways and frequent co-mineralization have made PGMs economically and strategically important as a group, especially considering their critical roles in industrial applications and their limited global distribution. Therefore, the descriptor "platinum-like" not only encompasses their chemical and physical similarities but also highlights their geological kinship and the collective approach required for their exploration, extraction, and processing.

THE DISCOVERY OF PLATINUM AND PLATINUM-GROUP METALS

Platinum itself was likely first encountered by pre-Columbian civilizations in South America, where indigenous peoples used it in alloys with gold to craft ornamental artifacts. However, the metal was not recognized as distinct by European scientists until the 18th century. When Spanish explorers encountered it in placer deposits in Colombia, they called it "*platina*," meaning "little silver," due to its silvery appearance and its presence alongside gold, mistakenly considering it an impurity in gold mining. It was not until 1748 that Antonio de Ulloa, a Spanish scientist and explorer, described platinum in detail, marking the first formal recognition of the metal in scientific literature (Scott and Bray, 1980). Around the same time, Charles Wood, an English metallurgist, independently identified and investigated platinum's unique properties (Hunt, 1985).

The discovery of palladium and rhodium is credited to William Hyde Wollaston, an English chemist who isolated these metals in the early 19th century while refining platinum ore from South America (Wollaston, 1809). In 1803, Wollaston announced the discovery of palladium, naming it after the asteroid Pallas, which had been discovered two years earlier. A year later, he isolated rhodium, naming it after the Greek word rhodon (ῥόδον), meaning "rose", due to the rose-colored salts it forms. Wollaston's work was groundbreaking, as he developed methods to purify and characterize these metals, paving the way for their future applications.

Ruthenium, iridium, and osmium were identified around the same time as palladium and rhodium, during investigations of the residues left after dissolving platinum ore in aqua regia. Iridium and osmium were discovered in 1803 by Smithson Tennant, an English chemist, who noted their remarkable properties in the insoluble residues from platinum processing (Lewis, 2011). Iridium was named after the Latin word iris (rainbow) due to the colorful nature of its salts, while osmium derives its name from the Greek word osme (smell) because of the strong odor of its volatile oxide. Ruthenium, however, was discovered slightly later, in 1844, by Karl Ernst Claus, a Russian chemist, who isolated it from platinum ores found in the Ural Mountains. He named it after "Ruthenia," the Latin name for Russia, in honor of its geographic origin (Seddon, 1996).

The identification and isolation of these metals were made possible by the development of advanced chemical techniques, particularly the use of aqua regia, a highly corrosive and powerful mixture of concentrated nitric acid and hydrochloric acid, to dissolve platinum-group ores that are otherwise resistant to most acids.

Platinum and PGMs belong to the d-block of the periodic table and are characterized by partially filled d-orbitals, which play a central role in their chemical behavior. For instance, platinum has an electron configuration of [Xe] $4f^{14}$ $5d^9$ $6s^1$, while palladium, with its completely filled d-orbital, has the configuration [Kr] $4d^{10}$, making it chemically distinct among the group. Other PGMs, such as rhodium ([Kr] $4d^8$ $5s^1$), ruthenium ([Kr] $4d^7$ $5s^1$), iridium ([Xe] $4f^{14}$ $5d^7$ $6s^2$), and osmium ([Xe] $4f^{14}$ $5d^6$ $6s^2$), exhibit a range of d-electron configurations that contribute to their ability to engage in complex chemical interactions.

The electronic structures of PGMs allow for multiple oxidation states, which vary depending on the specific element and its chemical environment. Platinum and palladium, for example, commonly exhibit oxidation states of 0, +2, and +4, while ruthenium and osmium can reach higher oxidation states, up to +8, reflecting their higher reactivity. These variable oxidation states are a key factor in the catalytic activity of PGMs, enabling them to participate in redox reactions and facilitate chemical transformations (Jeong et al., 2020). Their high ionization energies and strong relativistic effects, especially in heavier elements like platinum, iridium, and osmium, contribute to their chemical inertness and high stability under a wide range of conditions.

These metals are also known for their reluctance to form simple ionic compounds, instead favoring the formation of coordination complexes. This is due to the stability provided by their d-electrons, which interact strongly with ligands such as chlorides, phosphines, and amines. These complexes are critical in catalytic processes, as they enable PGMs to interact with small molecules like hydrogen, oxygen, and carbon monoxide, facilitating bond activation and formation. Additionally, the relativistic contraction of their orbitals enhances their density, compact atomic radii, and bonding strength, further reinforcing their chemical resilience and effectiveness as catalysts (Das et al., 2023).

PLATINUM AND PLATINUM-GROUP METAL EXTRACTION AND PROCESSING

Despite their numerous advantages, PGMs are relatively scarce in the Earth's crust. They are often found in low concentrations and are typically mined in specific regions, particularly in South Africa, Russia, and Canada. Unlike more abundant metals like iron or aluminum, they do not form large, easily accessible ore deposits. Instead, they are often dispersed in trace amounts within other minerals or ores, requiring specialized techniques to extract them. This rarity, coupled with their high demand, significantly contributes to the high market value of platinum and its group metals.

Currently, Russia and the Republic of South Africa are widely recognized as the leading global producers of PGMs. This is evidenced by production data from 2018, which shows that 84% of the world's platinum, palladium, and rhodium production occurs in these two countries (Matthey, 2019). In South Africa, the primary deposits of platinum-group metal (PGM) ores are found in the Bushveld Complex, located in the northern part of the country. The upper zone of this complex contains the highest concentrations of PGMs. The complex is composed of several key ore bodies, including the Upper Group 2 (UG-2) chromite ores, and the Merensky and Platreef ore beds. These reefs vary in grain size, PGM associations, and concentration levels. The Platreef ore body, based on current consumption rates, is estimated to have enough platinum reserves to last approximately 800 years. The western limestone deposits, which are rich in platinum-bearing ores, are mainly mined using underground techniques, while the eastern deposits are mined through open-pit methods, which are more cost-effective. As a result, many companies have shifted to open-pit mining in recent years due to its lower operational costs (Aleksandrova and O'Connor, 2020).

Russia is one of the world's largest producers of PGMs, with the majority of production being palladium, found predominantly in the sulfide copper-nickel ores of the Norilsk region and the Kola Peninsula, where 99.5% of the country's active PGM reserves are located. While Russia's platinum production is significantly smaller than that of South Africa, there are plans to increase platinum production by processing non-conventional platinum-bearing sources. These include ores from basic and ultrabasic ore bodies in regions such as the Urals, Aldan, Kamchatka, as well as carbon-rich black shales in the Far East and vanadium-iron-titanium formations in Karelia. Of particular interest are the low-grade ores from the South-Siberian Continental Plateau, which contain stratified ultrabasic rocks with potential for platinum recovery. Key Russian deposits include as Kondyor, Chernogorskoye, and the southern part of Norilsk-1. Kondyor, located in the Khabarovsk region, the world's largest known alluvial platinum deposit. In 2019, Kondyor produced 249.1 kg of placer platinum and 206.1 kg of chemically pure platinum (Aleksandrova and O'Connor, 2020).

This concentration of PGMs supplies creates geopolitical and economic risks for industries that depend heavily on these metals. For instance, any disruptions in the supply chain due to political instability, environmental regulations, or market fluctuations can lead to significant price increases and supply shortages. Given that demand for PGMs continues to rise, especially in emerging technologies like clean energy and advanced electronics, concerns about their future availability are growing. These concerns are compounded by the limited recycling capacity for these metals, accounting for about of 30% of global platinum and 29% of global palladium supplies in 2010 (Wilburn, 2012), which means that extraction from ores remains the primary PGM source. However, the growing recognition of the importance of PGMs has led to increased efforts in recycling, particularly in the automotive and electronics industries, with used catalytic converters, electronic waste, and other sources increasingly being recycled to recover these valuable metals, helping to ensure a more sustainable and circular economy (Xia and Ghahreman, 2023).

PLATINUM AND PLATINUM-GROUP METAL APPLICATIONS

Platinum and platinum-group metals are also critical to a wide range of modern technological and industrial applications, owing to their unique chemical stability, high melting points, and excellent catalytic properties, as well as remarkable versatility and efficiency in various high-demand processes.

One of the most significant applications of PGMs is in catalysis, a process where these metals act as catalysts to speed up chemical reactions without being consumed in the process. Platinum, palladium, and rhodium are especially important in automotive catalytic converters, where they help reduce harmful emissions, such as nitrogen oxides, carbon monoxide, and unburned hydrocarbons, from internal combustion engines. This catalytic function is crucial for meeting stringent environmental regulations and mitigating the impact of automotive pollution. PGMs also play a central role in chemical manufacturing, including the production of fertilizers, plastics, and synthetic fuels. For instance, platinum is essential in processes like the hydrogenation of oils and the refining of petroleum, transforming raw materials into valuable products. Without these metals, many of the chemical processes that fuel industrial economies would be far less efficient or even impossible.

The role of PGMs in energy systems is growing rapidly as the world transitions toward cleaner energy sources. Hydrogen fuel cells are one of the most prominent examples of how these metals are essential in the renewable energy sector. In fuel cells, platinum and palladium act as catalysts in the electrochemical reactions that generate electricity by combining hydrogen and oxygen, with water as the only byproduct. This technology holds significant potential for reducing carbon emissions, especially in the transportation sector, where hydrogen-powered vehicles are being explored as an alternative to fossil fuel-based engines. Additionally, PGMs are critical in hydrogen production technologies, such as electrolysis, which separates hydrogen from water using electricity. This process is seen as a key enabler of a sustainable hydrogen economy, providing cleaner fuel for diverse applications, from power generation to industrial use. Moreover, PGMs are being investigated for use in batteries and supercapacitors, which are key components of energy storage systems crucial for the integration of renewable energy sources like solar and wind.

In electronics, PGMs are highly valued for their excellent conductivity and resistance to corrosion, making them ideal materials for high-performance devices. They are used in critical components such as connectors, resistors, and capacitors, where their stability and efficiency are paramount for the reliability of devices in consumer electronics, telecommunications, and aerospace technologies. Furthermore, PGMs are being explored in the realm of nanotechnology, where their unique properties enable advancements in sensors, medical devices, and drug delivery systems. Platinum-based nanoparticles, for example, are being investigated for use in targeted cancer therapies, where their size and chemical reactivity allow them to interact with cells in ways that bulk platinum cannot.

Moreover, PGMs are seen as a critical component in the transition to a low-carbon economy. As countries and industries look for alternatives to fossil fuels, these metals are at the heart of technologies that aim to reduce greenhouse gas emissions and improve energy efficiency. The development of green technologies, such as hydrogen-powered vehicles, renewable energy storage, and sustainable chemical processes, relies on the continued availability of PGMs. Consequently, research into material substitutes for PGMs is ongoing, with some promising alternatives being explored. For example, investigations into the use of less rare metals, such as nickel or copper, in catalytic applications are being conducted, though these alternatives often do not yet match the performance and efficiency of PGMs. In addition, in some instances, substitution of PGMs has already taken place, and additional substitution may not be feasible. Moreover, since PGMs and their most promising substitutes are often co-produced from the same mineral deposits, their ability to replace each other during supply disruptions is restricted (Nassar, 2015). These factors indicate that the potential for substituting PGMs in current high-volume applications is limited. Therefore, further efforts are necessary to develop new technologies that minimize or eliminate the use of PGMs, reducing our dependence on them in the future. Table 4.1 summarizes some of the most important PGM applications.

Table 4.1 Some of the most important platinum and platinum-group metal applications summarizing their role and importance in different application areas

Application Area	PGMs Involved	Role and Importance
Catalysis (Automotive)	Platinum, Palladium, Rhodium	PGMs act as catalysts in automotive catalytic converters to reduce harmful emissions (nitrogen oxides, carbon monoxide, unburned hydrocarbons), essential for meeting environmental standards.
Chemical Manufacturing	Platinum, Palladium, Rhodium, Ruthenium, Iridium	PGMs facilitate chemical reactions in the production of fertilizers, plastics, synthetic fuels, hydrogenation of oils, and petroleum refining, making industrial processes more efficient.

Application Area	PGMs Involved	Role and Importance
Energy Systems (Fuel Cells)	Platinum, Palladium	In hydrogen fuel cells, PGMs catalyze the electrochemical reaction between hydrogen and oxygen to produce electricity, with water as the only byproduct, enabling cleaner energy production.
Energy Systems (Hydrogen Production)	Platinum, Palladium, Rhodium	PGMs play a vital role in hydrogen production technologies (e.g., electrolysis), which separate hydrogen from water, crucial for a sustainable hydrogen economy and low-carbon energy systems.
Energy Storage (Batteries and Supercapacitors)	Platinum, Palladium	PGMs are explored for use in energy storage systems, enhancing the performance of batteries and supercapacitors for renewable energy integration (solar, wind) and efficient energy storage.
Electronics	Platinum, Palladium, Rhodium, Iridium	PGMs are used in electronic components (connectors, resistors, capacitors) for their high conductivity, corrosion resistance, and stability in high-performance devices like consumer electronics.
Nanotechnology	Platinum, Palladium	PGMs are utilized in nanoparticle form for medical devices, sensors, and targeted cancer therapies, leveraging their unique chemical reactivity for precision applications.
Low-Carbon Economy Technologies	Platinum, Palladium, Rhodium, Iridium	PGMs are central to green technologies, such as hydrogen-powered vehicles and renewable energy storage, critical for reducing greenhouse gas emissions and advancing energy efficiency.
Research into Substitutes	Nickel, Copper (Alternative Metals)	Ongoing research into replacing PGMs with more abundant metals like nickel or copper for catalytic applications, although these alternatives are not yet as efficient as PGMs.

PLATINUM AND PLATINUM-GROUP METALS IN AQUATIC ECOSYSTEMS

Platinum and platinum-group metal applications are increasingly gaining attention due to their presence in the environment and their potential toxic effects on wildlife. These metals are often released into ecosystems through industrial processes, vehicle emissions, and the wear and tear of catalytic converters in automobiles. While these metals are considered relatively non-toxic to humans at low concentrations, their effects on wildlife can be more pronounced due to bioaccumulation and toxicity at elevated levels.

Despite the increasing recognition of PGMs in the environment, mainly emitted by automotive catalytic converters and hospital effluents containing Pt-based anti-cancer drugs, the available data on their distribution is still sparse, and further studies are required to better understand their fate and ecological impacts. Furthermore, the high persistence of PGMs in the environment means that once they enter aquatic ecosystems, they can remain for extended periods, continuously impacting local wildlife. Although limited in number, some studies have explored the distribution of PGMs in various environmental compartments such as water, particulate matter, and sediments across different regions worldwide. In addition, several assessments have highlighted the toxic effects of PGMs on wildlife, particularly in aquatic environments. Exposure to platinum and PGMs has been shown to cause different deleterious effects, such as oxidative stress, cellular damage, altered hormone levels and reduced survival, with potential impacts on growth, reproduction, and behavior (Ravindra et al., 2004; Pawlak et al., 2014).

One study in this sense investigated the distribution and biogeochemical processes controlling Pt in coastal systems, focusing on Genoa Harbor (Italy), by determining dissolved platinum (PtD) and particulate platinum (PtP), along with variables such as dissolved oxygen, organic carbon, chlorophyll-a, turbidity, and ammonium levels (Abdou et al., 2019a). The reported findings indicate distinct Pt concentration gradients linked to various sources like hospital, domestic, and industrial wastewater, as well as road runoff. Concentrations up to 0.18 ng L^{-1} for PtD and 14 ng g^{-1} for PtP indicate significant anthropogenic contamination in the harbor. The findings suggest that Pt could serve as a new tracer for human pollution in urbanized coastal systems, emphasizing the importance of detailed environmental monitoring to understand Pt's partitioning and fate.

Another study investigated the distribution of Pt in the pristine Krka River, in Croatia, estuary during winter and summer (Pađan et al., 2019). The results indicate that Pt concentrations in surface waters increased towards the seawater end-member, while bottom waters showed higher Pt concentrations near the freshwater end-member, likely due to accumulation over time. The study also noted that 90% of the Pt was in dissolved form. A slight deviation from the expected mixing pattern at lower salinities suggested potential changes in Pt redox speciation. The findings highlight the spatial variations of Pt in estuarine environments and emphasize the importance of understanding Pt behavior in such ecosystems.

Another assessment examined Pt contamination in the northwestern Mediterranean coast, focusing on temporal and spatial variations (Abdou et al., 2019b). The study reported historical Pt sources in Toulon Bay sediments, including those predating vehicle emission devices, such as the petroleum industry and coal-fired activities. The authors indicate that a significant increase in Pt concentrations in both sediments and farmed mussels is noted over the past 25 years, reflecting the rise in Pt demand, particularly for vehicle catalysts. The study also assessed Pt levels in wild mussels along a ~700 km stretch of coastline in the study area, revealing variations in Pt exposure across different ecosystems. The findings emphasize the need for ongoing monitoring of Pt levels in coastal areas to evaluate its environmental impact and potential risks to marine life.

Another assessment study examined short-term variations in platinum (Pt) concentrations and primary production indicators across three contrasting coastal sites during the spring bloom: Gironde Estuary mouth (SW France), Arcachon Bay (SW France), and Genoa Harbor (NW Italy) (Abdou et al., 2020a). The study compared diel cycles of dissolved Pt concentrations and primary production indicators (chlorophyll-a, phaeopigments, and particulate organic carbon) in surface seawater, plankton, and bivalves. In the Gironde Estuary, phytoplankton activity was found to influence Pt cycling, with zooplankton grazing potentially releasing Pt into the dissolved phase at night. External factors like tides or local Pt sources were more prominent in urbanized and confined areas such as Arcachon Bay and Genoa Harbor. The results suggest that Pt contamination is common in urbanized coastal waters, and Pt transfer to the marine food chain begins with primary producers. Phytoplankton, by concentrating Pt, may serve as an effective bioindicator of Pt contamination in coastal environments.

A study of dissolved Pt concentrations in eleven large pristine river systems across East Asia found that most samples had concentrations below 1.4 pmol L^{-1}, with a small fraction reaching up to 5.8 pmol L^{-1} (Soyol-Erdene and Huh, 2012). Platinum concentrations varied slightly between rivers, with the highest median concentration found in the Huang He (0.63 pmol L^{-1}) and the lowest in the Duman (0.18 pmol L^{-1}). The primary difference in Pt yield was due to variations in runoff. Rivers draining the eastern Tibetan Plateau, such as the Salween, Mekong, and Yangtze, exhibited higher Pt yields than rivers in the Russian Far East, like the Amur and Lena. Extrapolating the average riverine Pt concentration globally, the estimated riverine flux of Pt to the ocean is 13×10^3 mol y^{-1}, with an oceanic residence time of approximately 24 ± 10 thousand years.

Two assessments have investigated Pt concentrations in coastal waters of Japan, after the 2011 tsunami. The first investigated concentrations between the Boso and Sanriku areas (Mashio et al., 2017). Pt concentrations were found to range from 0.20 to 1.5 pmol L^{-1}, with the highest levels in bottom waters near the Boso coastal area and

Tokyo Bay. The study suggests that Pt exists in the form of $PtCl_5(OH)^{2-}$, even in low-oxygen waters. Vertical profiles showed an increase in Pt levels towards the seafloor near coastal stations, with concentrations resembling those of the open ocean at trench stations. High Pt levels are believed to be sourced from coastal sediments. The second assessment determined Pt concentrations in seawater and sediment in Otsuchi Bay, Japan (Mashio et al., 2020). The concentrations of dissolved Pt in seawater ranged from 0.40 to 1.99 pmol L^{-1}, while river water had concentrations below the detection limit. Pt levels in sediment samples varied from 0.46 to 14.4 ng g^{-1}. Seasonal variations showed that Pt concentrations in seawater increased with depth during the stratification season, likely due to Pt released from the sediments. Post-depositional processes, such as the remobilization of loosely adsorbed Pt on sediment particles, contributed to the increased Pt input in the bay, with 1.3–5.6% of the leachable fraction released, potentially transported by the tsunami.

Another study evaluated Pt contamination in estuarine and coastal sediments from the Tagus Estuary and Prodelta shelf (Cobelo-García et al., 2011). Elevated Pt concentrations were found in the upper layers of estuarine sediments, with up to 25 times the background levels, particularly near vehicular traffic sources, reaching a maximum of 9.5 ng g^{-1}. Concentrations decreased with depth, indicating no significant contamination before the use of catalytic converters in automobiles. In the Tagus Prodelta shelf, a sub-surface maximum in Pt levels suggested potential post-depositional mobility, which may obscure the anthropogenic Pt contamination signature in coastal sediments. That study, thus, highlights the impact of traffic-related pollution on Pt levels in coastal areas.

Another sediment assessment evaluated the concentrations of Pt, Rh and Pd in the Santos-São Vicente region of Brazil, which is heavily impacted by industrial activity, highway traffic, and sewage outfalls (Berbel et al., 2021). The region, home to the largest harbor in Latin America and an industrial hub, shows high levels of contamination, with Rh concentrations ranging from 0.08 to 1.7 ng g^{-1}, Pt from 0.15 to 40.3 ng g^{-1}, and Pd from 1.05 to 22.0 ng g^{-1}. The highest Rh concentrations were found near domestic waste, Pt was highest near ferryboat traffic, and Pd levels exceeded 5 ng g^{-1} in 50% of the stations. Spatial distribution of the Platinum Group Elements (PGEs) did not correlate with sediment types, as high levels were found in both muddy and sandy sediments. Pollution indexes such as anthropogenic factor (AF), geoaccumulation index (Igeo), Enrichment factor (EF), and Pollution Load Index (PLI) indicated significant to strong PGE contamination at 50% of the sediment samples. Stations near the Santos, São Vicente, and Bertioga channels exhibited high anthropogenic enrichment. However, no correlations were found between the PGE elements, though typical auto-catalyst ratios were observed at two stations near highways, suggesting a contribution from road dust and traffic-related emissions. The results highlight the complex biogeochemical cycling of PGEs in urbanized estuarine and coastal waters.

In another study, the biogeochemical behavior of Pt and Pd were assessed in coastal sediments, with and without the presence of the deposit-feeding invertebrate *Arenicola marina* to improve the understanding of Pt and Pd availability, cycling, and fate in contaminated coastal ecosystems (French and Turner, 2008). When metals were introduced as solutes, *A. marina* enhanced their sorption to sediment by exposing more particle surface area to contaminated water. After 10 days, assimilation efficiency (AE) for Pt was about 10%, and for Pd, it was around 1%. Calculations showed that both aqueous and dietary metal sources were significant. When Pt and Pd were introduced via ground catalytic converter particles, *A. marina* caused more metal solubilization, with AE values of less than 0.1% for Pt and around 1% for Pd. Table 4.2 summarizes the main findings of these assessments.

Several laboratory-based experiments have been conducted to assess how PGMs are absorbed, compartmentalized, and excreted by aquatic organisms, aiming to understand the bioavailability, bioaccumulation, and toxicological effects of these elements in different aquatic species. Various models, including freshwater and marine organisms like fish, mollusks, and algae, have been used to study the uptake of PGMs from contaminated water or sediment.

One study, for example, investigated the uptake of Rh, Pd and Pt by the marine macroalga *Ulva lactuca* (Turner et al., 2007). The removal of these metals from filtered seawater followed pseudo-first-order kinetics, with different rate constants depending on whether the reaction was irreversible or reversible. The equilibrium distribution coefficients for Rh, Pd, and Pt were found to be approximately 1,400, 900, and 350 mL g^{-1}, respectively. Uptake of Rh significantly increased with higher seawater pH, while Pt removal showed a small increase, and Pd uptake was not influenced by pH. The internalization of metals within cells was about 40% for Rh, 80% for Pd, and 95% for Pt. The study suggests that Rh is adsorbed to the algal surface as cationic hydrated chloride complexes, Pd binds strongly to organic ligands, and Pt requires specific interactions with physiologically active sites for internalization.

Another study on algae aimed to understand the marine biogeochemistry of Rh, Pd and Pt by exposing the marine microalga *Chlorella stigmatophora* to ppb concentrations of these metals in seawater (Shams et al., 2014). In short-term (24-h) exposures, the extent of metal accumulation followed the order: Rh > Pd >> Pt, with Pd being internalized to a greater extent than Rh and Pt. Accumulation was linear up to 30 µg L^{-1}, and higher algal biomass led to reduced

Table 4.2 Main findings concerning platinum group metal distributions in aquatic ecosystems and their environmental compartments

Study	Organism	Location	Analyzed Metal(s)	Medium (Sediment/ Seawater)	Findings/Comments
French and Turner (2008)	*Arenicola marina*	Coastal sediments	Pt, Pd	Sediment	A. marina enhanced metal sorption; Pt assimilation efficiency (AE) ~10%, Pd AE ~1%; lower AE when metals introduced via catalytic converter particles.
Cobelo-García et al. (2011)	–	Tagus Estuary, Prodelta shelf, Portugal	Pt	Sediment	Elevated Pt levels in estuarine sediments, especially near vehicular traffic sources; post-depositional mobility in Prodelta shelf.
Soyol-Erdene and Huh (2012)	–	East Asia (Rivers)	Pt	Seawater	Pt concentrations below 1.4 pmol L^{-1}; variation in Pt yield due to runoff; global Pt flux estimate to the ocean.
Mashio et al. (2017)	–	Boso and Sanriku, Japan	Pt	Seawater	Pt concentrations from 0.20 to 1.5 pmol L^{-1}; higher levels near Boso and Tokyo Bay; vertical profiles show increased Pt near seafloor.
Abdou et al. (2019b)	Mussels	Toulon Bay, NW Mediterranean	Pt	Sediment/ Seawater	Increase in Pt in sediments and mussels over 25 years; reflects rise in Pt demand for vehicle catalysts.
Abdou et al. (2019a)	–	Genoa Harbor, Italy	PtD, PtP	Seawater	Pt concentration gradients linked to sources like wastewater and road runoff; Pt could be a new tracer for human pollution.
Abdou et al. (2020a)	Phytoplankton, zooplankton, bivalves	Gironde Estuary, Arcachon Bay, Genoa Harbor	Pt	Seawater	Pt contamination linked to primary producers; urbanized coastal waters show Pt transfer to food chain.
Mashio et al. (2020)	–	Otsuchi Bay, Japan	Pt	Sediment/ Seawater	Pt concentrations 0.40 to 1.99 pmol L^{-1} in seawater; sediment concentrations 0.46 to 14.4 ng g^{-1}; seasonal variations in Pt release.
Pađan et al. (2019)	–	Krka River Estuary, Croatia	Pt	Seawater	Pt concentrations higher at seawater end-member; 90% of Pt in dissolved form; variations in redox speciation.
Berbel et al. (2021)	–	Santos-São Vicente, Brazil	Pt, Rh, Pd	Sediment	High Pt, Rh, Pd levels; significant contamination due to industrial activity, traffic, and sewage; no correlation with sediment types.

metal accumulation due to the production of exudates that stabilize metals in the water. In longer-term exposures (156-h), Rh showed the highest degree of accumulation and internalization, while Pd and Pt had similar accumulation levels, influenced by kinetic constraints. The study suggests that Rh plays a significant role in biological removal and transport processes in the marine environment, while Pd and Pt may behave differently in transient processes.

Another study investigated the toxicity of platinum (Pt), palladium (Pd), and rhodium (Rh) to the aquatic organism *Daphnia magna* under standardized conditions (Zimmermann et al., 2017). The toxicity was tested in single-metal exposure experiments following OECD guideline 202, measuring immobility and lethality at 24 and 48 hours. The results revealed that Pd was the most toxic, followed by Pt and Rh, with 48-hour LC50 values of 14 µg L^{-1} for Pd, 157 µg L^{-1} for Pt, and 56,800 µg L^{-1} for Rh. The toxicity was more pronounced for Pt and Rh in terms of immobility, while Pd showed similar toxicity for both immobility and lethality. The study also found synergistic toxicity when Pt and Pd were exposed together, indicating a more-than-additive effect. However, the authors indicate that the modes of action behind this synergistic effect remain unclear.

Several assessments have been conducted regarding PGMs in bivalves. For example, one study examined the distribution and biological availability of Pt in European river systems, where automobile emissions contribute to metal pollution. The distribution of Pt and other traffic-related metals (Cd, Cr, Cu, Ni, Pb, Zn) was analyzed in sediment samples and the Asian clam *Corbicula sp.* collected from three transects near road runoff inlets (Ruchter and Sures, 2015). The highest Pt concentrations were found in the silt/clay fraction of sediments, and Pt concentrations were correlated with the size of the drained street section and sediment grain size. Although the clams accumulated Pt, the concentrations in their soft tissue remained low (maximum 1.3 ng/g), suggesting that acute toxic effects are unlikely. However, potential chronic exposure effects require further investigation.

In another assessment, the uptake and bioaccumulation of 15 road dust metals, including traffic-related PGMs Pd, Pt and Rh by zebra mussels (*Dreissena polymorpha*) was evaluated in laboratory exposure studies (Zimmermann et al., 2002). Mussels were maintained in two types of water: nonchlorinated tap water and humic water from a bog lake, both containing dust from a moderately frequented road. After 26 weeks of exposure, the mussel soft tissues were analyzed for metal concentrations. Results showed that mussels bioaccumulated metals such as Pd, Ag, Bi, Cd, Co, Cr, Cu, Fe, Pb, and Sb, with concentrations ranging from several hundred micrograms per gram for some metals (e.g., Fe) to less than 10 ng g^{-1} for PGEs. The bioaccumulation factors

(BAF) of metals decreased in the following order: Cu > Cd > Ag > Pd > Sb > Pb > Fe > Pt > Rh. Humic water enhanced the biological availability of most metals compared to tap water. The study revealed an unrecognized high availability of Pd in the environment, suggesting that Pd should be more intensively monitored to assess its distribution in the biosphere.

One study investigated the accumulation and trophic transfer of Rh, Pd, and Pt in the marine macroalga *Ulva lactuca* and the grazing mollusc *Littorina littorea* over a 5-day exposure period (Mulholland and Turner, 2011). The metals were added to seawater at concentrations of 20 µg/L. *U. lactuca* accumulated metals in the order Rh, Pt > Pd, while *L. littorea*accumulated them in the order Pd ≥ Pt ≥ Rh, with the highest metal accumulation occurring in the visceral complex and kidney. When fed contaminated algae, *L. littorea* showed an order of magnitude increase in Rh and Pd accumulation, but Pt was not readily detected. The study concluded that the diet is the primary vector for Rh and Pd accumulation in *L. littorea*, while Pt is primarily accumulated from the aqueous phase. Grazing molluscs like *L. littorea* could serve as effective biomonitors for coastal PGE contamination.

Another study investigated the ecotoxicological effects of cisplatin (CDDP), a widely used chemotherapy drug containing Pt, on the marine mussel *Mytilus galloprovincialis*. Mussels were exposed to CDDP at a concentration of 100 ng/L for 14 days (Trombini et al., 2016). The study assessed various biomarkers, including antioxidant enzyme activities (superoxide dismutase, catalase, glutathione peroxidase), phase II enzyme activity (glutathione-S-transferase), oxidative damage (lipid peroxidation), genotoxicity (DNA damage), and neurotoxicity (acetylcholinesterase activity). The results revealed that CDDP exposure led to significant changes in the antioxidant capacity, oxidative stress in target organs (digestive gland and gills), DNA damage in hemocytes, and neurotoxicity. The authors highlight the potential risk of cisplatin exposure to non-target aquatic organisms, even at low concentrations commonly found in the aquatic environment.

In another study, also conducted in bivalves, examined the induction of heat shock proteins (hsp70) in zebra mussels (*Dreissena polymorpha*) following exposure to (Pt), (Pd), and (Rh), comparing them to the effects of cadmium (Cd) and lead (Pb) (Singer et al., 2005). Mussels were exposed for 10 weeks, and their metal concentrations and hsp70 levels were analyzed weekly. Results showed the highest metal uptake for Cd, followed by Pt, Pb, and Pd, with Rh showing the lowest uptake. Hsp70 levels increased over time in all exposed groups, with maximum concentrations reached by day 39, after which they decreased. The threshold metal levels for hsp70 induction varied, with the order of sensitivity being Rh < Pd ≤ Pb < Pt < Cd. Pd induced the highest increase in hsp70 levels, followed by Pt and Rh, which were 19 times higher than controls. Cd and Pb showed much lower hsp70 induction. This study highlights the significant cellular effects of PGEs, especially Pd, which had the lowest threshold and strongest hsp70 induction among the metals tested.

One study investigated the accumulation of Pt in oysters (*Crassostrea gigas*) exposed to varying concentrations of the stable isotope ^{194}Pt in seawater (Abdou et al., 2018). The authors indicate that, despite its growing presence in the environment, Pt levels remain very low, making detection challenging. Oysters were exposed to three Pt concentrations (50, 100, and 10,000 ng L^{-1}) over 35 days, with regular seawater renewal. The results showed a concentration-dependent accumulation of Pt in oysters, with lower exposure levels reaching a plateau after 28 days, while higher exposure levels continued to increase. The study found a linear correlation between seawater Pt concentrations and oyster accumulation for low and intermediate exposure levels, suggesting that oysters could be effective sentinels for monitoring Pt contamination in coastal marine environments. In another assessment by the same research group, the physiological effects of the bioconcentration of Pt in oysters (*Crassostrea gigas*) exposed to different concentrations of Pt (50, 100, and 10,000 ng L^{-1}) for 35 days were observed (Abdou et al., 2020b). The oysters accumulated Pt mainly in the gills and, to a lesser extent, in the mantle, with a time- and concentration-dependent increase in Pt levels across various organs (digestive gland, gonads, gills, mantle, and muscle). At the highest concentration (10,000 ng L^{-1}), significant cellular impairments were observed, including altered lipid contents and metal accumulation in lysosomes, suggesting activation of detoxification mechanisms, such as Pt excretion through feces. Despite no significant changes in the condition index, the integrative biological response (IBR) index indicated a decline in the overall health of the oysters, emphasizing the importance of the digestive gland in the organism's response to Pt exposure.

Another study examined the bioavailability and bioaccumulation of platinum chloride in the model organism zebrafish (*Danio rerio*) and in ramshorn snails (*Marisa cornuarietis*), focusing on the effects of platinum on embryonic development (Osterauer et al., 2009). The tested concentrations ranged from environmentally relevant levels (38 ng L^{-1}) to higher concentrations (up to 74.2 µg L^{-1} for zebrafish and 98.7 µg L^{-1} for snails). Both organisms accumulated platinum, with bioaccumulation factors (BAFs) ranging from 5–55 for zebrafish and 218.4–723.9 for snails. Platinum exposure altered heart rates at the lowest concentrations and, at higher concentrations, reduced hatching rates, delayed development, and caused weight loss in snails. Another study by the same research group investigated the effects of different concentrations of platinum

($PtCl_2$) on the same organisms, focusing on the induction of heat shock protein 70 (hsp70) and histopathological alterations (Osterauer et al., 2010). The results showed that Pt exposure led to histopathological changes in both species, with the severity varying depending on the Pt concentration. In snails, hsp70 levels did not increase, while in zebrafish, a significant rise in hsp70 was observed at the highest Pt concentration (100 $\mu g\ L^{-1}$). These findings highlight the potential biological effects of platinum, especially at higher concentrations.

One study focused on the bioaccumulation of Pt and Rh in European eels (*Anguilla anguilla*), exposed to Pt^{4+} and Rh^{3+} in tap and humic water over six weeks (Zimmermann et al., 2004). The results showed significant accumulation of Pt (68–840 ng g^{-1}) and Rh (35–155 ng g^{-1}) in the bile, liver, kidney, and intestine, with control groups showing negligible PGM levels. Rhodium was also detected in the gills and spleen, while no uptake occurred in the muscle or blood. The distribution of these metals varied between Pt and Rh and was influenced by water type. Due to their capacity to accumulate PGM, eels were, thus, identified by the authors as effective indicators for detecting PGM pollution in aquatic ecosystems.

Another study investigated the bioaccumulation of Pt, Pd and Rh by the freshwater isopod *Asellus aquaticus* in both natural ecosystems and laboratory conditions (Moldovan et al., 2001). The isopods collected from an urban river exhibited the following mean concentrations: 155.4 ng g^{-1} for Pd, 38.0 ngg^{-1} for Pt, and 17.9 ngg^{-1} for Rh. In laboratory conditions, *A. aquaticus* exposed to PGE standard solutions for 24 hours showed bioaccumulation factors of 150 for Pd, 85 for Pt, and 7 for Rh. The study found that bioaccumulation is time-dependent and increases with higher PGE concentrations in the environment. Additionally, the uptake rate of the three PGMs differed depending on whether the isopods were exposed to catalyst materials or environmental samples, likely due to transformations of PGMs species in the environment.

Another study investigated the distribution of PGMs, particularly rhodium (Rh) and platinum (Pt), in the water, macroalgae, and mussels of two urbanized estuarine systems (Douro and Ave estuaries, Portugal) (Abdou et al., 2023b). The study found that PGM contamination in these marine organisms was linked to urban effluent sources, with Pt/Rh ratios indicating that traffic emissions significantly impacted the Douro estuary, while industrial inputs influenced the Ave estuary. The concentrations of PGMs in macroalgae and mussels reflected urban environmental pressures and suggested their potential as bioindicators for PGM contamination in estuarine and coastal areas. Table 4.3 summarizes the findings reported by these field-based assessments.

Table 4.3 Main findings concerning field-bases platinum group metal assessments

Study	Organism	Location	Analyzed Metals	Main Findings
Moldovan et al. (2001)	*Asellus aquaticus* (freshwater isopod)	Urban river	Pd, Pt, Rh	Bioaccumulation factors differed; Pd showed highest accumulation, followed by Pt and Rh.
Zimmermann et al. (2002)	*Dreissena polymorpha* (zebra mussels)	–	Pd, Pt, Rh	Mussels bioaccumulated Pd, Pt, and Rh; humic water enhanced metal bioavailability.
Zimmermann et al. (2004)	*Anguilla anguilla* (European eel)	–	Pt, Rh	Eels accumulated Pt and Rh, primarily in bile, liver, and kidney; eels as bioindicators for PGM pollution.
Singer et al. (2005)	*Dreissena polymorpha* (zebra mussels)	–	Pt, Pd, Rh, Cd, Pb	Pt induced highest increase in hsp70 levels, followed by Pd and Rh; lower threshold levels for hsp70 induction in PGEs.
Turner et al. (2007)	*Ulva lactuca* (macroalga)	–	Rh, Pd, Pt	Uptake of Rh, Pd, and Pt followed pseudo-first-order kinetics, with different rate constants; pH influenced Rh and Pd uptake.
Osterauer et al. (2009)	*Danio rerio* (zebrafish), *Marisa cornuarietis* (ramshorn snail)	–	Pt	Both species bioaccumulated Pt; higher concentrations led to developmental and weight loss effects in snails.
Osterauer et al. (2010)	*Danio rerio* (zebrafish), *Marisa cornuarietis* (ramshorn snail)	–	Pt	Histopathological changes and increased hsp70 in zebrafish at high Pt concentrations.
Mulholland and Turner (2011)	*Ulva lactuca* (macroalga), *Littorina littorea* (mollusc)	–	Rh, Pd, Pt	Metals accumulated in U. lactuca in order: Rh, Pt > Pd; L. littorea showed different accumulation patterns.
Shams et al. (2014)	*Chlorella stigmatophora* (microalga)	–	Rh, Pd, Pt	Rh accumulated the most, followed by Pd and Pt; higher algal biomass reduced metal accumulation.

Table 4.3 (Contd.)

Table 4.3 (*Contd.*) Main findings concerning field-bases platinum group metal assessments

Study	Organism	Location	Analyzed Metals	Main Findings
Ruchter and Sures (2015)	*Corbicula* sp. (clam)	European rivers	Pt, Cd, Cr, Cu, Ni, Pb, Zn	Pt concentrations highest in silt/clay fraction of sediment; low accumulation in clams, suggesting low acute toxicity.
Trombini et al. (2016)	*Mytilus galloprovincialis* (mussel)	–	Pt (cisplatin)	Significant changes in antioxidant capacity, oxidative stress, DNA damage, and neurotoxicity after Pt exposure.
Zimmermann et al. (2017)	*Daphnia magna*	–	Pt, Pd, Rh	Pd was the most toxic, followed by Pt and Rh; synergistic toxicity observed for Pt and Pd.
Abdou et al. (2018)	*Crassostrea gigas* (oyster)	–	Pt	Pt accumulation in oysters correlated with seawater concentrations; effective as sentinels for Pt monitoring.
Abdou et al. (2020b)	*Crassostrea gigas* (oyster)	–	Pt	Time- and concentration-dependent Pt accumulation in oysters; significant cellular impairments at high concentrations.
Abdou et al. (2023b)	Macroalgae and mussels	Douro and Ave estuaries, Portugal	Pt, Rh	PGE contamination linked to urban effluent sources; macroalgae and mussels as bioindicators for PGE contamination.

Chapter 5

Specific Technology-Critical Elements Issues

TECHNOLOGY-CRITICAL ELEMENTS AND MICROPLASTICS

Studies on the association between TCEs and micro- and nanoplastics are still relatively limited, especially when compared to other areas of environmental pollution. While plastic pollution has received substantial attention in recent years, much of the research on the interactions between plastics and environmental contaminants has predominantly focused on metals that are more commonly recognized for their toxicity or widespread environmental presence. Elements such as cadmium (Cd), mercury (Hg), and lead (Pb) are often prioritized due to their well-documented harmful effects on both ecosystems and human health, which has led to extensive studies exploring their bioaccumulation, toxicological mechanisms, and environmental persistence.

Among the scarce assessments carried out in this sense, on study assessed the occurrence of microplastic pellets (2–5 mm) and their composition in terms of trace metals and rare earth elements (REE) on nine sandy beaches along the coast of Bahia, northeastern Brazil (Souza et al., 2022). The concentrations of total REE in the pellets ranged from 0.36 to 1.74 mg kg^{-1}, with white/transparent pellets exhibiting approximately five times higher REE concentrations than brown pellets. Additionally, the sum of other trace metals (Fe, Cd, Cu, Ni, Pb, and Zn) was highest in white pellets (357 ± 12 mg kg^{-1}). This suggests that plastic pellets may serve as a significant vector for the exposure of biota to REE and trace metals. The authors note that, while the concentrations of REE

DOI: 10.1201/9781003412328-05

and trace metals measured in this study were not particularly high, a potential role of pellets as vectors for the transport and exposure of REEs and other trace metals is noted, contributing to growing concerns over the environmental impact of microplastics.

In another study, Rb was identified among various other elements present in microfibers found in oyster farming areas (Saldaña-Serrano et al., 2022). The microplastics were all smaller than 5 mm, comprising fiber lines, suggesting a possible link to effluent discharge from nearby urbanized regions. A Scanning Electron Microscopy analysis indicated degradation of the microplastics, potentially leading to the formation of nanoplastics. The authors indicate that the presence of Rb, along with other elements such as titanium, aluminum, and copper, suggests that the microfibers may also carry organic compounds.

Concerning PGMs, the sorption behavior of Pt and Pd on polyethylene microplastics (PE-MP) original aged in Milli-Q water, and aged in lake water was investigated in one study (Sajkowska and Leśniewska, 2024). The sorption of Pt(IV) and Pd(II) on the original PE-MP at natural water pH showed low efficiency due to the forms of the analytes and the surface charge of the PE-MP; however, water acidification enhanced this process. Aged PE-MP in natural lake water, subjected to solar radiation, exhibited surface degradation and biofilm formation. The sorption process followed pseudo-second-order kinetics, indicating that chemisorption dominates. A good fit with both the Freundlich and Langmuir isotherm models suggested a mixed nature of sorption. The high Freundlich constant indicated multilayer adsorption on the heterogeneous surface of PE-MP aged in lake water, while monolayer sorption was better described by the Langmuir model for PE-MP aged in Milli-Q water. Comparing sorption capacities for Pt(IV) and Pd(II) between original and aged PE-MP in lake water showed that surface characteristics of the microplastics, rather than the forms of the analytes, played a greater role in the amount of sorbed metals. These findings clearly demonstrate that Pt(IV) and Pd(II) can be preconcentrated on the surface of inert PE microplastics, even from waters with low concentrations, posing potential risks by transferring these elements over long distances, with implications for ecological and human health.

Also concerning PGMs, another study investigated the role of microplastics (MPs) as Pd carriers in the Mediterranean mussel, *Mytilus galloprovincialis* (Rodrigues et al., 2023). Mussels were exposed to Pd in two forms, namely water-dissolved and MPs-adsorbed, with subsequent depuration over 144 h. The results demonstrated that mussels exposed to MPs-adsorbed Pd accumulated 2–3 times higher Pd concentrations (59% of initial Pd) compared to those exposed to dissolved Pd (25%). The accumulation was significantly higher in the digestive gland when exposed to MPs-adsorbed Pd, while the gills accumulated more Pd from the water-dissolved form. Additionally, exposure to Pd induced oxidative stress and altered feeding behavior in the mussels. Pd exposure also caused an imbalance in immune functions, as shown by the disruption of bactericidal and protease activities, which could lead to increased susceptibility to infections. Surprisingly, Pd depuration from mussels was not facilitated by MPs, as observed in previous studies, indicating that Pd desorbed from MPs during digestion rather than being eliminated with feces. These findings suggest that MPs enhance Pd bioaccumulation and toxicity, and they underline the ecological risks posed by the increasing levels of this emerging pollutant in marine environments.

In one study on PGM-derived pharmaceuticals, the transport behavior of three platinum-based anticancer drugs cisplatin, oxaliplatin, and carboplatin, was assessed through soil–water environments, both with and without the presence of surface-functionalized polystyrene nanoparticles (PS-NPs, nanoplastics) (Nath et al., 2020). Carboplatin, when tested alone, showed "tracer-like" mobility through soil. However, the addition of aminated PS-NPs to the aqueous solutions reduced the mobility of carboplatin in soil. The transport of Pt-complexes originating from cisplatin alone resulted in an elution of 35% of the initial concentration at the early stages of the experiment, which gradually decreased to 15–20%. In contrast, the presence of carboxylated PS-NPs significantly increased the recovery of cisplatin Pt-complexes to approximately 56–60%. Oxaliplatin demonstrated the lowest mobility, with a recovery of only 5–10%. However, the addition of aminated PS-NPs increased the recovery by more than four-fold, reaching 35–36%.

TECHNOLOGY-CRITICAL ELEMENTS AND CLIMATE CHANGE

The ongoing climate change scenario will have both direct and indirect effects on the water environment, impacting currents, pollutant transformation, transport, and their toxic effects. Global warming and extreme weather events will, for example, exacerbate the ecological risks of contaminants in the aquatic environment, with anticipated increases in leaching, diffuse pollution, and sediment contamination release into the water column significantly intensifying these risks (Wu et al., 2017). The potential changes in TCE toxicodynamics following climate change highlight the need for further research into their potential risks in aquatic organisms, especially considering the likely alterations in environmental conditions driven by climate change. For example, understanding interactions between these elements and microplastics in different temperature regimes is essential for assessing long-term ecological consequences, as microplastics can act as vectors for pollutant transport, amplifying their effects on marine life (Rochman et al., 2013). Baseline studies are also urgently required in this regard, to provide data on current TCE contamination levels, uptake, and toxicological impacts, serving as reference points, enabling future comparisons with altered environmental scenarios driven by climate change, and ensuring that potential ecological risks are understood in the context of evolving environmental conditions.

In this sense, rising temperatures and altered precipitation patterns can accelerate weathering processes, influencing the geochemical behavior and mobility of TCEs in the environment, and potentially increasing metal mobilization (Rue and McKnight, 2021). In mining regions affected by climate change, for example, higher temperatures may lead to increased leaching of several TCEs from mining waste and tailings into surrounding ecosystems, contributing to environmental contamination (Anawar, 2013). One study conducted in Snake River, in the western USA, verified, for example that REE concentrations of archived samples have risen over time in parallel with increasing summer air temperatures (Rue and McKnight, 2021).

In particular, several elements, are known to be rapidly depurated at higher temperatures (Campbell et al., 2005). Studies have, for example, shown decreased Rb levels in bony fish tissues, *i.e.*, liver and peripheral tissues, at elevated temperatures simulating climate change (Barbosa et al., 2017; Peters et al., 1999). Decreases in Rb levels could disrupt ionic homeostasis, potentially impairing the normal functioning of K-dependent processes such as nerve signaling, muscle contraction, and osmoregulation, although further studies are required, as Rb's biological role is not yet well understood. On the other hand, higher temperatures can also stimulate increased sodium (Na) metabolism and respiratory activity in certain fish species, which may lead to higher Rb uptake (Gashkina and Moiseenko, 2020). Additionally, climate-induced changes in water chemistry, such as altered pH levels or salinity, may influence Rb bioavailability and uptake mechanisms in aquatic organisms, further complicating the understanding of its toxicological impact.

Another study, although not focusing on climate change, assessed the content of macro- and microelements, including Rb in the lower reaches of the Ob River (Western Siberia) over a four-year period (2020–2023), comparing seasons, *i.e.*, winter low water, spring-summer floods, and early fall low water, at three river stations (Moskovchenko et al., 2024). The results revealed significant seasonal variations in the elemental content, which were influenced by changes in catchment inputs, physical and chemical conditions of the aquatic environment, and the amount and composition of incoming suspended sediment.

During high water flow events in the Ob River, the concentration of suspended solids increased substantially, including elevated quantities of lithophilic elements such as Rb, Ti, Zr, Ga, and others. Specifically, the ratio between suspended and dissolved forms of elements like Rb varied seasonally. In winter, when the Ob River was ice-covered, a 2- to 3-fold rise in the concentration of Na, Mg, Ca, K, Si, Mn, and Rb was observed, with Rb predominantly found in the dissolved form. This contrasts with the flood periods, during which elements like Rb, along with other lithophilic elements, exhibited higher concentrations in the suspended sediment. The removal of hard hydrolysable elements, including Rb, mainly occurred during floods as part of the suspended load. The findings highlight the influence of seasonal hydrological changes on the transport and behavior of Rb and other elements in the river systems, with higher levels of Rb in the dissolved form during the winter and an increased presence in suspended solids during flood periods.

In another study, climate-driven changes, including rising air temperatures, enhanced permafrost degradation, and altered precipitation patterns, were found to have significant effects on biogeochemical cycling, particularly regarding Rb in two lakes, East Lake and West Lake, affected differently by climate change, located on Melville Island in the High Arctic (Muir et al., 2023). West Lake experienced substantial permafrost degradation, leading to a dramatic increase in turbidity due to multiple underwater mass movement events that began in 2008, significantly increasing the concentrations of bioaccumulative elements, including Rb. In particular, Rb concentrations rose by 1.4% per year between 2009 and 2017, reflecting the broader environmental changes occurring in the lake due to climate change. In contrast, East Lake, which experienced only minor permafrost degradation and no significant increase in turbidity, saw a decline in the concentrations of these elements, including a 5.6% annual reduction in Rb. This difference between the two lakes demonstrates how climate-induced factors such as permafrost thaw and changes in turbidity can alter the cycling of elements like Rb in Arctic ecosystems.

A similar assessment was conducted in Stordalen Mire, a peatland ecosystem located 200 km north of the Arctic Circle in northern Sweden, which was investigated to understand the impact of rapid permafrost thaw on elemental cycling, particularly Rb strontium (Sr) (Guillot-Wallace et al., 2023). That mire is undergoing significant changes, with thawing permafrost converting drier, raised permafrost palsas into saturated wetlands, including bogs and fens. As permafrost thaws, the hydrologic connectivity of the mire is altered, leading to changes in groundwater flow paths. The study aimed to use pore-water collected from the degrading peatland ecosystem for elemental analysis of Rb and Sr to improve understanding of the changing hydrologic flow paths, which have implications for carbon transport in the region. Surface and porewater samples were collected across the Stordalen Mire watershed and along a permafrost thaw transect over a 4-week period, including areas near the thaw front and a vegetated riparian zone. The samples were processed using trace element clean techniques to ensure accurate measurements. The results indicated that Rb/Sr ratios could track groundwater paths within the mire, providing insight into the shifting hydrological dynamics due to thawing. The comparison of sampling dates and transect locations suggested evidence of a dry period within the mire, which was corroborated by field observations of low water tables, low rainfall, and warm days. The study found that Rb/Sr ratios were useful for identifying when thaw ponds had become hydrologically connected. The authors state that these assessments can offer insights into the changing hydrology of the Arctic as a whole.

Concerning climate change effects, little is known about organism responses exposed to pollutants and increased temperature conditions, particularly concerning emerging contaminants like TCEs (Andrade et al., 2022). In this sense, one study investigated the impact of temperature on the effects of the La in the mussel *Mytilus galloprovincialis* (Andrade et al., 2022). Mussels were exposed to 10 µg L^{-1} of La at two temperatures (17°C and 22°C) for 28 days. Results showed that temperature did not affect La bioaccumulation, but La exposure reduced metabolic capacity and increased biotransformation enzyme activity as a defense mechanism. However, antioxidant defenses were inhibited, leading to higher lipid peroxidation (LPO) levels. Warming alone caused metabolic shutdown, reducing enzyme activities and protein carbonylation (PC) levels. Combined exposure to La and warming caused metabolic depression, activated biotransformation defenses, reduced antioxidant capacity, and increased cellular damage.

Another assessment examined the impact of warming La exposure on glass eels (*Anguilla anguilla*) exposed to 1.5 µg L^{-1} of La for five days, with an additional five days of depuration, at both present-day and +4°C warming scenarios (Figueiredo et al., 2020). The results showed that warming increased the bioaccumulation and toxicity of La, particularly in the viscera, followed by the head and body. Elimination of La was less effective at higher temperatures. Exposure to La did not affect acetylcholinesterase activity, but lipid peroxidation was significantly higher when both La and warming were combined. The expression of heat shock proteins was suppressed in glass eels exposed to La, indicating an inability to effectively prevent cellular damage, especially under warming conditions.

Ocean acidification effect assessments in TCE-exposure scenarios are also scarce. One assessment in this regard evaluated the combined effects of ocean warming, acidification, and La accumulation on the surf clam (*Spisula solida*) (Figueiredo et al., 2022b). Specimens were exposed to these stressors for 7 days and then depurated for an additional 7 days. The study found that La bioaccumulated rapidly, even under control and climate change scenarios, and could not be fully eliminated during the depuration phase. Biochemical responses were triggered, including increased levels of antioxidant enzymes like SOD, CAT, GST, and TAC, but lipoperoxidation indicated insufficient detoxification. The heat shock protein (HSP) response was inhibited in La treatments combined with warming and acidification. The highest levels of lipoperoxidation were observed in clams exposed to all three stressors, highlighting the toxic effects of La, particularly in the context of climate change.

Chapter **6**

Conclusions

A comprehensive overview of several TCEs is given herein, from their historical discovery, extraction processes from environmental matrices, current broad range of applications, and environmental and health impact assessments. Reports indicate varying trends in TCE accumulation across different environmental compartments, with sediment and water concentrations often serving as reservoirs of these contaminants, while biota can accumulate TCEs at higher levels. For example, concentrations of certain TCEs in sediment and water are frequently found to be lower compared to those in the tissues of aquatic organisms, where bioaccumulation factors (BAFs) can exceed 1,000-fold for metals and metalloids, depending on the species and the environmental conditions. This highlights the significant role that organisms play in the bioaccumulation of contaminants, especially in sediment-heavy ecosystems where uptake mechanisms are enhanced by particle adsorption. Aquatic organisms exhibit varied responses to TCE exposure, with factors such as temperature, water chemistry, and metabolic processes influencing their uptake and toxicity. Variations in water temperature, pH and salinity can significantly impact the speciation and mobility of TCEs, altering their bioavailability and, consequently, their toxicity to aquatic life. These complex interactions between environmental variables and TCE behavior in aquatic ecosystems necessitate further research to better understand their long-term impacts and mechanisms of action.

Several laboratory-based experiments have been conducted on the discussed TCEs, providing valuable insights into the molecular and cellular mechanisms underlying their toxicity. Field-based assessments, however, are still lacking when compared to other metals and metalloids.

DOI: 10.1201/9781003412328-06

Targeted research on the growing issue of microplastics in aquatic ecosystems is urgently needed to evaluate how microplastics influence the uptake, transport, and accumulation of TCEs in aquatic organisms, particularly in benthic species and filter-feeding organisms, which could amplify ecological risks. Finally, in the context of climate change, higher temperatures and extreme weather events may accelerate the mobilization and uptake of TCEs, potentially amplifying their ecological and toxicological impacts. Given these emerging trends, it is crucial to adopt a proactive approach in assessing the combined effects of climate change, microplastics, and TCE exposure to safeguard ecosystem health, prevent bioaccumulation and potential biomagnification in aquatic food chains.

ACKNOWLEDGEMENTS

Thanks are due to the Carlos Chagas Filho Foundation for Research Support of the State of Rio de Janeiro (FAPERJ) for financial support (Apoio ao Jovem Pesquisador Fluminense com Vínculo em ICTS do Estado do RJ 2021, grant number E-26/210.300/20222, Jovem Cientista do Nosso Estado 2021–2024 , grant number E-26/201.270/202, Programa Pensa Rio - Apoio ao estudo de temas relevantes e estratégicos para o estado do Rio de Janeiro 2025, grant number E-26/210.723/2025), as well as the Brazilian National Council of Scientific and Technological Development (productivity grants, process numbers 308811/2021-6 and 301422/2025-7). The implementation of the Projeto Pesquisa Marinha e Pesqueira is a compensatory measure established by the Conduct Adjustment Agreement under the responsibility of the PRIO company, conducted by the Federal Public Ministry-MPF/RJ. This work was also prepared with resources from the Judicial Agreement Term signed between the Federal Public Prosecutor's Office, the Public Prosecutor's Office of the State of Paraná and Petrobrás, with the intervention of the Brazilian Fund for Biodiversity (Funbio) and the Chico Mendes Institute for Biodiversity Conservation (ICMBio), within the scope of the Sentence Enforcement process No. 5001337-92.2012.4.04.7008, arising from Public Civil Actions 0000041-91.2010.404.7008 (PR) and No. 200270080002601, brought by the IAP -PARANÁ ENVIRONMENTAL INSTITUTE and by the MPPR and MPF, respectively against PETRÓ LEO BRASILEIRO S/A -PETROBRÁS.

References

Abdou, M., Dutruch, L., Schäfer, J., Zaldibar, B., Medrano, R., Izagirre, U., et al. 2018. Tracing platinum accumulation kinetics in oyster *Crassostrea gigas*, a sentinel species in coastal marine environments. Science of The Total Environment. 615: 652–663. https://doi.org/10.1016/j.scitotenv.2017.09.078.

Abdou, M., Schäfer, J., Gil-Díaz, T., Tercier-Waeber, M.-L., Catrouillet, C., Massa, F., et al. 2019a. Spatial variability and sources of platinum in a contaminated harbor—tracing coastal urban inputs. Environmental Chemistry. 17(2): 105–117. https://doi.org/10.1071/EN19160.

Abdou, M., Schäfer, J., Hu, R., Gil-Díaz, T., Garnier, C., Brach-Papa, C., et al. 2019b. Platinum in sediments and mussels from the northwestern Mediterranean coast: Temporal and spatial aspects. Chemosphere. 215: 783–792. https://doi.org/10.1016/j.chemosphere.2018.10.011.

Abdou, M., Zaldibar, B., Medrano, R., Schäfer, J., Izagirre, U., Dutruch, L., et al. 2020a. Organotropism and biomarker response in oyster Crassostrea gigas exposed to platinum in seawater. Environmental Science and Pollution Research. 27(4): 3584–3599. https://doi.org/10.1007/s11356-018-3443-7.

Abdou, M., Gil-Díaz, T., Schäfer, J., Catrouillet, C., Bossy, C., Dutruch, L., et al. 2020b. Short-term variations of platinum concentrations in contrasting coastal environments: The role of primary producers. Marine Chemistry. 222: 103782. https://doi.org/10.1016/j.marchem.2020.103782.

Abdou, M., Cobelo-Garcia, A., Santos-Echeandía, J. and Schäfer, J. 2023a. Chapter 11—Technology-critical elements. Contaminants of Emerging Concern in the Marine Environment. 401–438. https://doi.org/10.1016/B978-0-323-90297-7.00007-X.

Abdou, M., Monteiro, C.E., Brito, P., Neuparth, T., Pinheiro, M., Santos, M., et al. 2023b. Platinum Group Element distribution in water and marine biota from two impacted estuarine environments (Douro and Ave estuaries, Portugal). Marine Pollution Bulletin. 192: 114990. https://doi.org/10.1016/j.marpolbul.2023.114990.

Abrão, A. 1994. Química e tecnologia das terras-raras. Série tecnologia mineral. CETEM, Rio de Janeiro, 66:212.

Adunka, R. and Orna, M.V. 2018. Discoveries among the Rare Earths. Carl Auer von Welsbach: Chemist, Inventor, Entrepreneur. 17–34. https://doi.org/10.1007/978-3-319-77905-8_3.

Aharchaou, I., Beaubien, C., Campbell, P.G.C. and Fortin, C. 2020. Lanthanum and Cerium Toxicity to the Freshwater Green Alga *Chlorella fusca*: Applicability of the biotic ligand model. Environmental Toxicology and Chemistry. 39(5): 996–1005. https://doi.org/10.1002/etc.4707.

Ahearn, G.A., Mandal, P.K. and Mandal, A. 2004. Mechanisms of heavy-metal sequestration and detoxification in crustaceans: A review. Journal of Comparative Physiology B. 174(6): 439–452. https://doi.org/10.1007/s00360-004-0438-0.

Aide, M.T. and Aide, C. 2012. Rare Earth elements: Their importance in understanding soil genesis. https://doi.org/10.5402/2012/783876.

Aleksandrova, T.N. and O'Connor, C. 2020. Processing of platinum group metal ores in Russia and South Africa: current state and prospects. Записки Горного института. 244 (eng): 462–473.

Ali, S. and Katima, J. 2020. Technology Critical Elements and the GEF, A STAP Advisory Document. Scientific and Technical Advisory Panel to the Global Environment Facility. Washington, D.C.

Alibo, D.S. and Nozaki, Y. 1999. Rare earth elements in seawater: Particle association, shale-normalization, and ce oxidation. Geochimica et Cosmochimica Acta. 63(3): 363–372. https://doi.org/10.1016/S0016-7037(98) 00279-8.

Alkadi, H. 2020. A review on free radicals and antioxidants. Infectious Disorders —Drug Targets. 20(1): 16–26. https://doi.org/10.2174/187152651866618062812 4323.

Altuntaş, E., Ammon, J., Cahn, S.B. and DeMille, D. 2018. Measuring nuclear-spin-dependent parity violation with molecules: Experimental methods and analysis of systematic errors. Physical Review A. 97(4): 042101. https://doi.org/10.1103/PhysRevA.97.042101.

Amyot, M., Clayden, M.G., MacMillan, G.A., Perron, T. and Arscott-Gauvin, A. 2017. Fate and Trophic Transfer of Rare Earth Elements in Temperate Lake Food Webs. ACS Publications. https://doi.org/10.1021/acs.est.7b00739.

Anawar, H. Md. 2013. Impact of climate change on acid mine drainage generation and contaminant transport in water ecosystems of semi-arid and arid mining areas. Physics and Chemistry of the Earth, Parts A/B/C. 58–60: 13–21. https://doi.org/10.1016/j.pce.2013.04.002.

Andrade, M., Soares, A.M.V.M., Solé, M., Pereira, E. and Freitas, R. 2022. Do climate change related factors modify the response of *Mytilus galloprovincialis* to lanthanum? The case of temperature rise. Chemosphere. 307: 135577. https://doi.org/10.1016/j.chemosphere.2022.135577.

Anenburg, M., Broom-Fendley, S. and Chen, W. 2021. Formation of Rare Earth deposits in carbonatites. Elements: An International Magazine of Mineralogy, Geochemistry, and Petrology. 17(5): 327–332.

Arnold, W.D., Crouse, D.J. and Brown, K.B. 1965. Solvent extraction of cesium (and Rubidium) from ore liquors with substituted phenols. Industrial & Engineering Chemistry Process Design and Development. 4(3): 249–254.

Baar, H.J.W., Schijf, J. and Byrne, R.H. 1991. Solution chemistry of the rare earth elements in seawater. European Journal of Solid State and Inorganic Chemistry; (France). 28.

Babu, S.S., Prajith, A., Rao, V.P., Mohan, M.R., Ramana, R.V. and Satyasree, N. 2024. Sources of rare earths and prospects for a viable REE deposit in river sediments of Kerala, southwest India. EBSCOhost. 126(3): 345. https://doi.org/10.18520/cs/v126/i3/345-359.

Balaram, V. 2019. Rare earth elements: A review of applications, occurrence, exploration, analysis, recycling, and environmental impact. Geoscience Frontiers. 10(4): 1285–1303. https://doi.org/10.1016/j.gsf.2018.12.005.

Bandi, T.N. 2024. A Comprehensive Overview of Atomic Clocks and their Applications. Demo Journal. 1(1): 40–50.

Barbosa, V., Maulvault, A.L., Alves, R.N., Anacleto, P., Pousão-Ferreira, P., Carvalho, M.L., et al. 2017. Will seabass (*Dicentrarchus labrax*) quality change in a warmer ocean? Food Research International. 97: 27–36. https://doi.org/10.1016/j.foodres.2017.03.024

Barry, M.J. and Meehan, B.J. 2000. The acute and chronic toxicity of lanthanum to *Daphnia carinata*. Chemosphere. 41(10): 1669–1674. https://doi.org/10.1016/S0045-6535(00)00091-6.

Batapola, N.M., Dushyantha, N.P., Premasiri, H.M.R., Abeysinghe, A.M.K.B., Rohitha, L.P.S., Ratnayake, N.P., et al. 2020. A comparison of global rare earth element (REE) resources and their mineralogy with REE prospects in Sri Lanka. Journal of Asian Earth Sciences. 200: 104475. https://doi.org/10.1016/j.jseaes.2020.104475.

Batley, G.E. and Campbell, P.G.C. 2022. Metal contaminants of emerging concern in aquatic systems. Environmental Chemistry. 19(1): 23–40. https://doi.org/10.1071/EN22030.

Baumgart, A., Haluza, D., Prohaska, T., Trimmel, S., Pitha, U., Irrgeher, J., et al. 2024. In-use dissipation of technology-critical elements from vehicles and renewable energy technologies in Vienna, Austria: A public health matter? Journal of Industrial Ecology. 28(6): 1857–1870. https://doi.org/10.1111/jiec.13571.

Beauvieux, A., Fromentin, J.-M., Romero, D., Couffin, N., Brown, A., Metral, L., et al. 2024. Molecular fingerprint of gilthead seabream physiology in response to pollutant mixtures in the wild. Environmental Pollution. 340: 122789. https://doi.org/10.1016/j.envpol.2023.122789.

Berbel, G.B.B., Hortellani, M.A., Sarkis, J.E. de S., Chiozzini, V.G., Fávaro, D.I.T., Sutti, B.O., et al. 2021. Emerging contaminants (Rh, Pd, and Pt) in surface sediments from a Brazilian subtropical estuary influenced by anthropogenic activities. Marine Pollution Bulletin. 163: 111929. https://doi.org/10.1016/j.marpolbul.2020.111929.

BGS, British Geological Survey. 2011. Rare earth elements. Available at: https://nora.nerc.ac.uk/id/eprint/17448/1/Rare_Earth_Elements_profile%5B1%5D.pdf. Accessed on: 15 September 2025.

Bilewicz, A., Bartoś, B., Misiak, R. and Petelenz, B. 2006. Separation of ^{82}Sr from rubidium target for preparation of $^{82}Sr/^{82}Rb$ generator. Journal of Radioanalytical and Nuclear Chemistry. 268(3): 485–487. https://doi.org/10.1007/s10967-006-0195-5.

Binnemans, K., Jones, P.T., Blanpain, B., Van Gerven, T., Yang, Y., Walton, A., et al. 2013. Recycling of rare earths: A critical review. Journal of Cleaner Production. 51: 1–22. https://doi.org/10.1016/j.jclepro.2012.12.037.

Blaise, C., Gagné, F., Harwood, M., Quinn, B. and Hanana, H. 2018. Ecotoxicity responses of the freshwater cnidarian *Hydra attenuata* to 11 rare earth elements. Ecotoxicology and Environmental Safety. 163: 486–491. https://doi.org/10.1016/j.ecoenv.2018.07.033.

Blinova, I., Lukjanova, A., Muna, M., Vija, H. and Kahru, A. 2018. Evaluation of the potential hazard of lanthanides to freshwater microcrustaceans. Science of The Total Environment. 642: 1100–1107. https://doi.org/10.1016/j.scitotenv.2018.06.155.

Borch, T., Kretzschmar, R., Kappler, A., Cappellen, P.V., Ginder-Vogel, M., Voegelin, A., et al. 2010. Biogeochemical Redox Processes and their Impact on Contaminant Dynamics. Environmental Science & Technology. 44(1): 15–23. https://doi.org/10.1021/es9026248.

Bowen, R. 1994. Rubidium-Strontium Dating. Isotopes in the Earth Sciences. 162–200. https://doi.org/10.1007/978-94-009-2611-0_4.

Briant, N., Le Monier, P., Bruzac, S., Sireau, T., Araújo, D.F. and Grouhel, A. 2021. Rare earth element in Bivalves' soft tissues of French Metropolitan Coasts: Spatial and temporal distribution. Archives of Environmental Contamination and Toxicology. 81(4): 600–611. https://doi.org/10.1007/s00244-021-00821-7.

Brioschi, L., Steinmann, M., Lucot, E., Pierret, M.C., Stille, P., Prunier, J., et al. 2013. Transfer of Rare Earth Elements (REE) from natural soil to plant systems: implications for the environmental availability of anthropogenic REE. Plant and Soil. 366(1): 143–163. https://doi.org/10.1007/s11104-012-1407-0.

Bu-Olayan, A.H. and Thomas, B.V. 2020. Bourgeoning impact of the technology critical elements in the marine environment. Environmental Pollution. 265: 115064. https://doi.org/10.1016/j.envpol.2020.115064.

Campbell, L.M., Fisk, A.T., Wang, X., Köck, G. and Muir, D.C. 2005. Evidence for biomagnification of rubidium in freshwater and marine food webs. Canadian Journal of Fisheries and Aquatic Sciences. 62(5): 1161–1167. https://doi.org/10.1139/f05-027.

Canavese, C., DeCostanzi, E., Branciforte, L., Caropreso, A., Nonnato, A., Pietra, R., et al. 2001. Rubidium deficiency in dialysis patients. Journal of Nephrology. 14(3): 169–175.

Carpenter, D., Boutin, C., Allison, J.E., Parsons, J.L. and Ellis, D.M. 2015. Uptake and Effects of Six Rare Earth Elements (REEs) on Selected Native and Crop Species Growing in Contaminated Soils. PLOS ONE. 10(6): e0129936. https://doi.org/10.1371/journal.pone.0129936.

Castor, S.B. and Hedrick, J.B. 2006 Rare Earth Elements. Industrial Minerals Vol. 7th Ed. Society for Mining, Metallurgy, and Exploration, Littleton, Colorado, 769–792.

Cesbron, F.P. 1989. Mineralogy of the Rare-Earth Elements. Lanthanides, Tantalum and Niobium. 3–26. https://doi.org/10.1007/978-3-642-87262-4_1.

CGEE, Centro de Gestão e Estudos Estratégicos. 2013. Usos e aplicações de Terras Raras no Brasil: 2012–2030. Centro de Gestão e Estudos Estratégicos, Brasília, 258.

Chakhmouradian, A.R. and Zaitsev, A.N. 2012. Rare Earth Mineralization in Igneous Rocks: Sources and processes. Elements. 8(5): 347–353. https://doi.org/10.2113/gselements.8.5.347

Chen, M., Wu, S., Tan, Y., Li, R., Liu, Y. and Huang, Q. 2019. Rubidium-doped titanium surfaces with modulatory effects on MC3T3-E1 cell response and antibacterial capacity against *Staphylococcus aureus*. Biomedical Materials. 14(4): 045016. https://doi.org/10.1088/1748-605X/ab2585.

Chen, Q., He, Z., Zhuo, Y., Li, S., Yang, W., Hu, L., et al. 2021. Rubidium chloride modulated the fecal microbiota community in mice. BMC Microbiology. 21(1): 46. https://doi.org/10.1186/s12866-021-02095-4.

Chen, Z., Li, Z., Chen, J., Kallem, P., Banat, F. and Qiu, H. 2022. Recent advances in selective separation technologies of rare earth elements: A review. Journal of Environmental Chemical Engineering. 10(1): 107104. https://doi.org/10.1016/j.jece.2021.107104.

Chiasson, A.G. 1991. The Ratio of Rubidium to Caesium in Threespine Stickleback (*Gasterosteus aculeatus*), Benthic and Limnetic Sticklebacks (*Gasterosteus*), and Juvenile Sockeye Salmon (*Oncorhynchus nerka*). Chemistry and Ecology. 5(4): 227–240. https://doi.org/10.1080/02757549108035252.

Choi, Y., Ryu, S., Naidu, G., Lee, S. and Vigneswaran, S. 2019. Integrated submerged membrane distillation-adsorption system for rubidium recovery. Separation and Purification Technology. 218: 146–155. https://doi.org/10.1016/j.seppur.2019.02.050.

Cidu, R., Vittori Antisari, L., Biddau, R., Buscaroli, A., Carbone, S., Da Pelo, S., et al. 2013. Dynamics of rare earth elements in water–soil systems: The case study of the Pineta San Vitale (Ravenna, Italy). Geoderma. 193–194: 52–67. https://doi.org/10.1016/j.geoderma.2012.10.009.

Cobelo-García, A., Neira, P., Mil-Homens, M. and Caetano, M. 2011. Evaluation of the contamination of platinum in estuarine and coastal sediments (Tagus Estuary and Prodelta, Portugal). Marine Pollution Bulletin. 62(3): 646–650. https://doi.org/10.1016/j.marpolbul.2010.12.018.

Colburn, N.E., Gage, S.G. and Gillen, C.M. 2001. Rubidium uptake is increased in shore crabs, *Carcinus maenas*, acclimated to dilute seawater. Physiological and Biochemical Zoology. https://doi.org/10.1086/322926.

Commission on Isotopic Abundances and Atomic Weights. 2015. Rubidium. CIAAW. Available at: https://www.ciaaw.org/rubidium.htm. Accessed on: 28 November 2024.

Cornett, C.R., Ehmann, W.D., Wekstein, D.R. and Markesbery, W.R. 1998. Trace elements in Alzheimer's disease pituitary glands. Biological Trace Element Research. 62(1): 107–114. https://doi.org/10.1007/BF02820026.

Da Ros, Z., Dell'Anno, A., Morato, T., Sweetman, A.K., Carreiro-Silva, M., Smith, C.J., et al. 2019. The deep sea: The new frontier for ecological restoration. Marine Policy. 108: 103642. https://doi.org/10.1016/j.marpol.2019.103642.

Dai, W., Fang, Y., Yu, L., Zhao, G. and Yan, X. 2018. Rubidium ion capture with composite adsorbent PMA@HKUST-1. Journal of the Taiwan Institute of Chemical Engineers. 84: 222–228. https://doi.org/10.1016/j.jtice.2018.01.023.

Dang, D.H., Filella, M. and Omanović, D. 2021. Technology-Critical Elements: An emerging and vital resource that requires more in-depth investigation. Archives of Environmental Contamination and Toxicology. 81(4): 517–520. https://doi.org/10.1007/s00244-021-00892-6.

Das, A., Das, U., Das, R. and Das, A.K. 2023. Relativistic effects on the chemistry of heavier elements: why not given proper importance in chemistry education at the undergraduate and postgraduate level? Chemistry Teacher International. 5(4): 365–378. https://doi.org/10.1515/cti-2023-0043.

Davies, J.A. and Hartley, F.R. 1981. Complexes of the platinum metals containing weak donor ligands. ACS Publications. 81: 70–90. https://doi.org/10.1021/cr00041a004.

Deng, X.L., Lau, C.P., Lai, K., Cheung, K.F., Lau, G.K. and Li, G.R. 2007. Cell cycle-dependent expression of potassium channels and cell proliferation in rat mesenchymal stem cells from bone marrow. Cell Proliferation. 40(5): 656–670. https://doi.org/10.1111/j.1365-2184.2007.00458.x.

Deng, Y., Ren, J., Guo, Q., Cao, J., Wang, H. and Liu, C. 2017. Rare earth element geochemistry characteristics of seawater and porewater from deep sea in western Pacific. Scientific Reports. 7(1): 16539. https://doi.org/10.1038/s41598-017-16379-1.

Deng, P., Lin, K., Yuan, W., Gomez, M.A., She, J., Yu, S., et al. 2024. Risk assessment and strontium isotopic tracing of potentially toxic metals in creek sediments around a uranium mine, China. Chemosphere. 353: 141597. https://doi.org/10.1016/j.chemosphere.2024.141597.

Denton, K.M., Ponce, D.A., Peacock, J.R. and Miller, D.M. 2019. Geophysical characterization of a Proterozoic REE terrane at Mountain Pass, eastern Mojave Desert, California, USA. Geosphere. 16(1): 456–471. https://doi.org/10.1130/GES02066.1.

Diaz, L.A., Lister, T.E., Parkman, J.A. and Clark, G.G. 2016. Comprehensive process for the recovery of value and critical materials from electronic waste. Journal of Cleaner Production. 125: 236–244. https://doi.org/10.1016/j.jclepro.2016.03.061.

Dubé, M., Auclair, J., Hanana, H., Turcotte, P., Gagnon, C. and Gagné, F. 2019. Gene expression changes and toxicity of selected rare earth elements in rainbow trout juveniles. Comparative Biochemistry and Physiology Part C: Toxicology & Pharmacology. 223: 88–95. https://doi.org/10.1016/j.cbpc.2019.05.009.

Dushyantha, N., Batapola, N., Ilankoon, I.M.S.K., Rohitha, S., Premasiri, R., Abeysinghe, B., et al. 2020. The story of rare earth elements (REEs): Occurrences, global distribution, genesis, geology, mineralogy and global production. Ore Geology Reviews. 122: 103521. https://doi.org/10.1016/j.oregeorev.2020.103521.

Elderfield, H., Hawkesworth, C.J., Greaves, M.J. and Calvert, S.E. 1981. Rare earth element geochemistry of oceanic ferromanganese nodules and associated sediments. Geochimica et Cosmochimica Acta. 45(4): 513–528. https://doi.org/10.1016/0016-7037(81)90184-8.

El-Sheikh, A., Driggers, C.M. and Shyng, S.-L. 2024. Non-radioactive Rb+ Efflux assay for screening KATP channel modulators. Potassium Channels: Methods and Protocols. 191–210. https://doi.org/10.1007/978-1-0716-3818-7_12.

Evans, C.H. 2012. Episodes from the History of the Rare Earth Elements. Springer Science & Business Media. Kluwer Academic Publishers, Dordrecht, Netherlands, 268.

Feng, W., Deng, Y., Yang, F., Miao, Q. and Ngien, S.K. 2023. Systematic review of Contaminants of Emerging Concern (CECs): Distribution, risks, and implications for water quality and health. Water. 15(22): 3922. https://doi.org/10.3390/w15223922.

Fernandez, V. 2017. Rare-earth elements market: A historical and financial perspective. Resources Policy. 53: 26–45. https://doi.org/10.1016/j.resourpol.2017.05.010.

Figueiredo, C., Oliveira, R., Lopes, C., Brito, P., Caetano, M. and Raimundo, J. 2022a. Rare earth elements biomonitoring using the mussel *Mytilus galloprovincialis* in the Portuguese coast: Seasonal variations. Marine Pollution Bulletin. 175: 113335. https://doi.org/10.1016/j.marpolbul.2022.113335.

Figueiredo, C., Grilo, T.F., Oliveira, R., Ferreira, I.J., Gil, F., Lopes, C., et al. 2022b. Single and combined ecotoxicological effects of ocean warming, acidification and lanthanum exposure on the surf clam (*Spisula solida*). Chemosphere. 302: 134850. https://doi.org/10.1016/j.chemosphere.2022.134850.

Figueiredo, C., Raimundo, J., Lopes, A.R., Lopes, C., Rosa, N., Brito, P., et al. 2020. Warming enhances lanthanum accumulation and toxicity promoting cellular damage in glass eels (*Anguilla anguilla*). Environmental Research. 191: 110051. https://doi.org/10.1016/j.envres.2020.110051.

Filella, M. and Rodríguez-Murillo, J.C. 2017. Less-studied TCE: are their environmental concentrations increasing due to their use in new technologies? Chemosphere. 182: 605–616. https://doi.org/10.1016/j.chemosphere.2017.05.024.

Filella, M. 2020. TCEs and environmental research: is the TCEs concept scientifically fruitful? Environmental Science and Pollution Research. 27(16): 20565–20570. https://doi.org/10.1007/s11356-020-08621-7.

French, B. and Turner, A. 2008. Mobilization, Adsorption, and Bioavailability of Pt and Pd in Coastal Sediments: The Role of the Polychaete, Arenicola marina. ACS Publications. https://doi.org/10.1021/es071693n.

Furuta, H. and Ohtsu, M. 1989. Evaluations of frequency shift and stability in rubidium vapor stabilized semiconductor lasers. Applied Optics. 28(17): 3737–3743. https://doi.org/10.1364/AO.28.003737.

Gantayat, R.R. and Elumalai, V. 2024. Salinity-induced changes in heavy metal behavior and mobility in semi-arid coastal aquifers: A comprehensive review. Water. 16(7): 1052. https://doi.org/10.3390/w16071052.

Gashkina, N.A. and Moiseenko, T.I. 2020. Biogeochemical Migration of Elements within the Lake Watershed of a Reference Site (Valdai Upland). Doklady Earth Sciences. 491(2): 264–270. https://doi.org/10.1134/S1028334X20040042.

Gausepohl, F., Hennke, A., Schoening, T., Köser, K. and Greinert, J. 2020. Scars in the abyss: reconstructing sequence, location and temporal change of the 78 plough tracks of the 1989 DISCOL deep-sea disturbance experiment in the Peru Basin. Biogeosciences. 17(6): 1463–1493. https://doi.org/10.5194/bg-17-1463-2020.

Gergoric, M., Ravaux, C., Steenari, B.-M., Espegren, F. and Retegan, T. 2018. Leaching and Recovery of Rare-Earth Elements from Neodymium Magnet Waste Using Organic Acids. Metals. 8(9): 721. https://doi.org/10.3390/met8090721.

Giese, E.C., Magalhães, D.D.P. and Egler, S.G. 2016. Biossorção de elementos de terras-raras. Serie tecnologia mineral. CETEM, Rio de Janeiro. 90: 75.

González, V., Vignati, D.A.L., Pons, M.-N., Montarges-Pelletier, E., Bojic, C. and Giamberini, L. 2015. Lanthanide ecotoxicity: First attempt to measure environmental risk for aquatic organisms. Environmental Pollution. 199: 139–147. https://doi.org/10.1016/j.envpol.2015.01.020.

Greenwood, I.A. and Weston, A.H. 1993. Effects of rubidium on responses to potassium channel openers in rat isolated aorta. British Journal of Pharmacology. 109(4): 925–932. https://doi.org/10.1111/j.1476-5381.1993.tb13709.x.

Guillot-Wallace, C., Fahnestock, M.F., Drum, K., Becker, K., Ortega, M., Gomez Rodriguez, Y., et al. 2023. Hydrologic Connectivity of the Stordalen Mire Watershed: Rubidium and Strontium Elements as Tracers of Surface-Groundwater Interactions. AGU Fall Meeting Abstracts. B23H-218.

Guo, X., Fan, N., Liu, Y., Liu, X., Wang, Z., Xie, X., et al. 2023. Deep seabed mining: Frontiers in engineering geology and environment. International Journal of Coal Science & Technology. 10(1): 23. https://doi.org/10.1007/s40789-023-00580-x.

Gupta, C.K. and Krishnamurthy, N. 1992. Extractive metallurgy of rare earths. International Materials Reviews. 37(1): 197–248. https://doi.org/10.1179/imr.1992.37.1.197.

Hahn, O., Strassman, F., Mattauch, J. and Ewald, H. 1943. Geologische Altersbestimmungen mit der Strontiummethode. Chemische Zeitung. 67: 55–56.

Hakala, R.W. 1952. Letters. Journal of Chemical Education. 29: 1–2. https://doi.org/10.1021/ed029p581.2.

Halka, M. and Nordstrom, B. 2019. Metals and Metalloids, 2nd Ed. Facts On File, New York.

Hanana, H., Kleinert, C. and Gagné, F. 2021. Toxicity of representative mixture of five rare earth elements in juvenile rainbow trout (*Oncorhynchus mykiss*) juveniles. Environmental Science and Pollution Research. 28(22): 28263–28274. https://doi.org/10.1007/s11356-020-12218-5.

Harmer, R.E. and Nex, P.a.M. 2016. Rare Earth Deposits of Africa. Episodes Journal of International Geoscience. 39(2): 381–406. https://doi.org/10.18814/epiiugs/2016/v39i2/95784.

Hass, E.M. 2006. Staying Healthy with Nutrition, 21st Century Edition: The Complete Guide to Diet & Nutritional Medicine. Celestial Arts, Berkeley, CA, USA, 223–224.

Hauser-Davis, R.A., Lopes, A.P., Willmer, I.Q., Monteiro, F., Saint'Pierre, T.D., Vianna, M., et al. 2023. Baseline data on rubidium (Rb) contamination in juvenile caribbean sharpnose shark *Rhizoprionodon porosus* from southeastern Brazil. Arq. Ciencias do Mar. 56(1): 1–14.

Herrmann, H., Nolde, J., Berger, S. and Heise, S. 2016. Aquatic ecotoxicity of lanthanum – A review and an attempt to derive water and sediment quality criteria. Ecotoxicology and Environmental Safety. 124: 213–238. https://doi.org/10.1016/j.ecoenv.2015.09.033.

Hoffmann, L.J., Breitbarth, E., Boyd, P.W. and Hunter, K.A. 2012. Influence of ocean warming and acidification on trace metal biogeochemistry. Marine Ecology Progress Series. 470: 191–205. https://doi.org/10.3354/meps10082.

Hogstrand, C. and Haux, C. 1990. Metallothionein as an indicator of heavy-metal exposure in two subtropical fish species. Journal of Experimental Marine Biology and Ecology. 138(1): 69–84. https://doi.org/10.1016/0022-0981(90)90177-E.

Hopfe, S., Flemming, K., Lehmann, F., Möckel, R., Kutschke, S. and Pollmann, K. 2017. Leaching of rare earth elements from fluorescent powder using the tea fungus Kombucha. Waste Management. 62: 211–221. https://doi.org/10.1016/j.wasman.2017.02.005.

Hunt, L.B. 1985. The First Experiments on Platinum. Platinum Metals Review. 29: 180–184.

Ikemoto, T., Tu, N.P.C., Okuda, N., Iwata, A., Omori, K., Tanabe, S., et al. 2008. Biomagnification of trace elements in the Aquatic Food Web in the Mekong Delta, South Vietnam using Stable Carbon and Nitrogen Isotope analysis. Archives of Environmental Contamination and Toxicology. 54(3): 504–515. https://doi.org/10.1007/s00244-007-9058-5.

ISA, International Seabed Authority. 2024. About ISA. Available at: https://www.isa.org.jm/about-isa/.Accessed on: 28 November 2024.

Janiga, M., Janiga, M. and Pitoňáková, T. 2024. Differential accumulation of metals in the lacustrine and fluvial Alpine bullheads (*Cottus poecilopus*) and recovery of fish from metal contamination after a flash flood. Environmental Science and Pollution Research. 31(11): 17387–17400. https://doi.org/10.1007/s11356-024-32288-z.

Jeong, H., Shin, D., Kim, B.-S., Bae, J., Shin, S., Choe, C., et al. 2020. Controlling the Oxidation State of Pt Single Atoms for Maximizing Catalytic Activity. Angewandte Chemie (International Ed. in English). 59(46): 20691–20696. https://doi.org/10.1002/anie.202009776.

Jiang, D.G., Yang, J., Zhang, S. and Yang, D.J. 2012. A survey of 16 rare earth elements in the major foods in China. Biomedical and Environmental Sciences. 25(3): 267–271. https://doi.org/10.3967/0895-3988.2012.03.003.

Jin, H., Reed, D.W., Thompson, V.S., Fujita, Y., Jiao, Y., Crain-Zamora, M., et al. 2019. Sustainable Bioleaching of Rare Earth Elements from Industrial Waste Materials Using Agricultural Wastes. ACS Sustainable Chemistry & Engineering. 7(18): 15311–15319. https://doi.org/10.1021/acssuschemeng.9b02584.

Johansson, B. 2000. Structural and electronic relationships between the lanthanide and actinide elements. Hyperfine Interactions. 128(1): 41–66. https://doi.org/10.1023/A:1012667128586.

Jordens, A., Cheng, Y.P. and Waters, K.E. 2013. A review of the beneficiation of rare earth element bearing minerals. Minerals Engineering. 41: 97–114. https://doi.org/10.1016/j.mineng.2012.10.017.

Kang, S., Guo, C., Xue, C., Ma, C., Mu, H. and Sun, L. 2022. Toxic Effects of Two Representative Rare Earth Elements (La and Gd) on Danio rerio Based on Transcriptome Analysis. Toxics. 10(9): 519. https://doi.org/10.3390/toxics10090519.

Kapitola, J. and Vilimovska, D. 1984. Age dependence of bone tissue uptake of radioactive rubidium ^{86}Rb in rats. Sb. Lek.; (Czechoslovakia). 86(10): 298–302.

Kaster, T., Mylonas, I., Renaud, J.M., Wells, G.A., Beanlands, R.S.B. and deKemp, R.A. 2012. Accuracy of low-dose rubidium-82 myocardial perfusion imaging for detection of coronary artery disease using 3D PET and normal database interpretation. Journal of Nuclear Cardiology. 19(6): 1135–1145. https://doi.org/10.1007/s12350-012-9621-y.

Kazmi, S.S.U.H., Wang, Y.Y.L., Cai, Y.-E. and Wang, Z. 2022. Temperature effects in single or combined with chemicals to the aquatic organisms: An overview of thermo-chemical stress. Ecological Indicators. 143: 109354. https://doi.org/10.1016/j.ecolind.2022.109354.

Khan, A.M., Bakar, N.K.A., Bakar, A.F.A. and Ashraf, M.A. 2017. Chemical speciation and bioavailability of rare earth elements (REEs) in the ecosystem: a review. Environmental Science and Pollution Research. 24(29): 22764–22789. https://doi.org/10.1007/s11356-016-7427-1.

Kirchhoff, G. and Bunsen, R. 1860. Chemische Analyse durch Spectralbeobachtungen. Annalen der Physik. 186(6): 161–189. https://doi.org/10.1002/andp.18601860602.

Koch, E.C. 2002. Special materials in pyrotechnics Part 2. Application of cæsium and rubidium compounds in pyrotechnics. Journal of Pyrotechnics. 15: 9–15. Available at: http://www.jpyro.co.uk/wp-content/uploads/j15_09_uh556.pdf. Accessed on: 28 November 2024.

Kocsis, L., Vennemann, T.W., Ulianov, A. and Brunnschweiler, J.M. 2015. Characterizing the bull shark Carcharhinus leucas habitat in Fiji by the chemical and isotopic compositions of their teeth. Environmental Biology of Fishes. 98(6): 1609–1622. https://doi.org/10.1007/s10641-015-0386-4.

Kouhail, Y., Abdou, M. and Gil-Díaz, T. 2022. Technology Critical Elements in groundwater resources - knowledge and gaps in the early 2020's. Current

Opinion in Environmental Science and Health. 26: 100329. https://doi.org/10.1016/j.coesh.2022.100329.

Lachaux, N., Cossu-Leguille, C., Poirier, L., Gross, E.M. and Giamberini, L. 2022. Integrated environmental risk assessment of rare earth elements mixture on aquatic ecosystems. Frontiers in Environmental Science. 10. https://doi.org/10.3389/fenvs.2022.974191.

Lapido-Loureiro, F.E.V. 2013. O Brasil e a reglobalização da indústria das terras-raras. CETEM/MCTI, Brasil.

Leite, C., Russo, T., Pinto, J., Polese, G., Soares, A.M.V.M., Pretti, C., et al. 2024. From the cellular to tissue alterations induced by two rare earth elements in the mussel species *Mytilus galloprovincialis*: Comparison between exposure and recovery periods. Science of The Total Environment. 915: 169754. https://doi.org/10.1016/j.scitotenv.2023.169754.

Lewis, D.G. 2011. The Early Life of Smithson Tennant FRS (1761–1815). Platinum Metals Review. 55(3): 196–200. https://doi.org/10.1595/147106711X581710.

Li, B., Liu, H., Ye, X., Li, S. and Wu, Z. 2014. Rubidium and Cesium Ion Adsorption by a Potassium Titanium Silicate-Calcium Alginate Composite Adsorbent. Separation Science and Technology. 49(7): 1076–1085. https://doi.org/10.1080/01496395.2013.869230.

Li, Z., Pranolo, Y., Zhu, Z. and Cheng, C.Y. 2017. Solvent extraction of cesium and rubidium from brine solutions using 4-tert-butyl-2-(α-methylbenzyl)-phenol. Hydrometallurgy. 171: 1–7. https://doi.org/10.1016/j.hydromet.2017.03.007.

Lister, T.E., Diaz, L.A., Clark, G.G. and Keller, P. 2016. Process development for the recovery of critical materials from electronic waste (No. INL/CON-15-36506). Idaho National Lab. (INL), Idaho Falls, ID (United States).

Liu, S., Liu, H., Huang, Y. and Yang, W. 2015. Solvent extraction of rubidium and cesium from salt lake brine with t-BAMBP–kerosene solution. Transactions of Nonferrous Metals Society of China. 25(1): 329–334. https://doi.org/10.1016/S1003-6326(15)63608-1.

Lodge, M.W., Segerson, K. and Squires, D. 2017. Sharing and Preserving the Resources in the Deep Sea: Challenges for the International Seabed Authority. International Journal of Marine and Coastal Law. 32: 427.

Lombeck, I., Kasperek, K., Feinendegen, L.E. and Bremer, H.J. 1980. Rubidium—A possible essential trace element. Biological Trace Element Research. 2(3): 193–198. https://doi.org/10.1007/BF02785354.

Lu, Y., Long, X., Mao, Y., Wang, L., Liao, X. and Zhao, L. 2025. Bacteriostatic activity and mechanism of minerals containing rubidium. Science of Traditional Chinese Medicine. 3(2): 137–144. https://doi.org/10.1097/st9.0000000000000064.

Lv, Y., Xing, P., Ma, B., Liu, Y., Wang, C., Zhang, W., et al. 2020. Efficient extraction of lithium and rubidium from polylithionite via alkaline leaching combined with solvent extraction and precipitation. ACS Sustainable Chemistry and Engineering. 8(38): 14462–14470. https://doi.org/10.1021/acssuschemeng.0c04437.

Maât, N., Nachbaur, V., Lardé, R., Juraszek, J. and Breton, J.-M.L. 2016. An innovative process using only water and sodium chloride for recovering rare earth elements from Nd–Fe–B permanent magnets found in the waste of electrical and electronic equipment. ACS Publications. https://doi.org/10.1021/acssuschemeng.6b01226.

Macdougall, J.D. and Harriss, R.C. 1969. The geochemistry of an Arctic watershed. Canadian Journal of Earth Sciences. https://doi.org/10.1139/e69-026.

MacMillan, G.A., Chételat, J., Heath, J.P., Mickpegak, R. and Amyot, M. 2017. Rare earth elements in freshwater, marine, and terrestrial ecosystems in the eastern Canadian Arctic. Environmental Science: Processes and Impacts. 19(10): 1336–1345. https://doi.org/10.1039/C7EM00082K.

Madadi, R., Kachoueiyan, F. and De-la-Torre, G.E. 2023. Effect of redox potential on the heavy metals binding phases in estuarine sediment: Case study of the Musa Estuary. Marine Pollution Bulletin. 195: 115565. https://doi.org/10.1016/j.marpolbul.2023.115565.

Mahmoud, M.G., Salem, S.G., Lotfy, I.M., Dawoud, M.M., Abu-Elhasan, M., Elithy, M.A., et al. 2024. Ecological Risk Assessment of Heavy Metals Pollution in Recent Sediments of Qaroun Lake, Fayoum, Egypt. Egyptian Journal of Chemistry. 67(10): 487–501. https://doi.org/10.21608/ejchem.2024.303404.9989.

Martinez, R.E., Pourret, O., Faucon, M.-P. and Dian, C. 2018. Effect of rare earth elements on rice plant growth. Chemical Geology. 489: 28–37. https://doi.org/10.1016/j.chemgeo.2018.05.012.

Mashio, A.S., Obata, H. and Gamo, T. 2017. Dissolved Platinum Concentrations in Coastal Seawater: Boso to Sanriku Areas, Japan. Archives of Environmental Contamination and Toxicology. 73(2): 240–246. https://doi.org/10.1007/s00244-017-0373-1.

Mashio, A., Obata, H., Shimazaki, T., Fukuda, H. and Ogawa, H. 2020. Spatiotemporal variations of platinum in seawater in Otsuchi Bay, Japan after the 2011 tsunami. Science of The Total Environment. 708: 134659. https://doi.org/10.1016/j.scitotenv.2019.134659.

Massari, S. and Ruberti, M. 2013 Rare earth elements as critical raw materials: Focus on international markets and future strategies. Resources Policy. 38: 36–43. https://doi.org/10.1016/j.resourpol.2012.07.001

Matthey, J. 2019. PLC. PGM Market Report: February 2019 [Internet]. London: Johnson Matthey plc; 2019 Feb [cited 2025 Dec 22]. Available from: https://matthey.com/documents/161599/162993/pgm-market-report-february-2019.pdf/f558c574-e8d9-5d8e-fe38-6590eb0a53f3?t=1650968146348

Mavukkandy, M.O., Chabib, C.M., Mustafa, I., Al Ghaferi, A. and AlMarzooqi, F. 2019. Brine management in desalination industry: From waste to resources generation. Desalination. 472: 114187. https://doi.org/10.1016/j.desal.2019.114187.

Mayfield, D.B. and Fairbrother, A. 2015. Examination of rare earth element concentration patterns in freshwater fish tissues. Chemosphere. 120: 68–74. https://doi.org/10.1016/j.chemosphere.2014.06.010.

Mccord, M.Q. 2021. The Essential Role of Recycling in Decreasing Americas Dependence on Foreign Sources of Rare Earth Minerals. Master thesis of Military Studies, Marine Corps University.

Merschel, G. and Bau, M. 2015. Rare earth elements in the aragonitic shell of freshwater mussel *Corbicula fluminea* and the bioavailability of anthropogenic lanthanum, samarium and gadolinium in river water. Science of The Total Environment. 533: 91–101. https://doi.org/10.1016/j.scitotenv.2015.06.042.

Miao, S., DeLaune, R.D. and Jugsujinda, A. 2006. Influence of sediment redox conditions on release/solubility of metals and nutrients in a Louisiana Mississippi River deltaic plain freshwater lake. Science of The Total Environment. 371(1): 334–343. https://doi.org/10.1016/j.scitotenv.2006.07.027.

Milinovic, J., Rodrigues, F.J.L., Barriga, F.J.A.S. and Murton, B.J. 2021. Ocean-Floor Sediments as a Resource of Rare Earth Elements: An Overview of Recently Studied Sites. Minerals. 11(2): 142. https://doi.org/10.3390/min11020142.

Miranda Jr, P. 2000. Síntese, caracterização e estudo termoanalítico de picratos de lantanídeos hidratados, dos ligantes 1, 3-ditiano-1-óxido (DTSO), cis-1, 3-ditiano-1, 3-dióxido (cis-DTSO2) e trans-1, 3-ditiano-1, 3-dióxido (trans-DTSO2) e de seus compostos de adição. Doctoral dissertation, Universidade de São Paulo.

Moldovan, M., Rauch, S., Gómez, M., Antonia Palacios, M. and Morrison, G.M. 2001. Bioaccumulation of palladium, platinum and rhodium from urban particulates and sediments by the freshwater isopod *Asellus aquaticus*. Water Research. 35(17): 4175–4183. https://doi.org/10.1016/S0043-1354(01)00136-1.

Moskovchenko, D., Soromotin, A., Khoroshavin, V., Prikhodko, N., Kirillov, V., Koveshnikov, M., et al. 2024. Climate change and the Ob river: A reassessment of major and trace element fluxes to the Arctic Ocean. Water. 16(15): 2112. https://doi.org/10.3390/w16152112.

Muir, D., Hudelson, K., Burke, S., Kirk, J., Tetreault, G., Barst, B., et al. 2023. Divergent temporal trends of Mercury, Cesium, Rubidium, and Strontium in Arctic Char from paired lakes influenced by increased turbidity and other climate-related drivers. *In*: AGU Fall Meeting Abstracts. B31C-06.

Mulholland, R. and Turner, A. 2011. Accumulation of platinum group elements by the marine gastropod *Littorina littorea*. Environmental Pollution. 159(4): 977–982. https://doi.org/10.1016/j.envpol.2010.12.009.

Naidu, G., Nur, T., Loganathan, P., Kandasamy, J. and Vigneswaran, S. 2016a. Selective sorption of rubidium by potassium cobalt hexacyanoferrate. Separation and Purification Technology. 163: 238–246. https://doi.org/10.1016/j.seppur.2016.03.001.

Naidu, G., Loganathan, P., Jeong, S., Johir, M.A.H., To, V.H.P., Kandasamy, J., et al. 2016b. Rubidium extraction using an organic polymer encapsulated potassium copper hexacyanoferrate sorbent. Chemical Engineering Journal. 306: 31–42. https://doi.org/10.1016/j.cej.2016.07.038.

Naidu, G., Jeong, S., Choi, Y., Song, M.H., Oyunchuluun, U. and Vigneswaran, S. 2018. Valuable rubidium extraction from potassium reduced seawater brine. Journal of Cleaner Production. 174: 1079–1088. https://doi.org/10.1016/j.jclepro.2017.11.042.

Nassar, N.T. 2015. Limitations to elemental substitution as exemplified by the platinum-group metals. Green Chemistry. 17(4): 2226–2235. https://doi.org/10.1039/C4GC02197E.

Nath, J., Dror, I. and Berkowitz, B. 2020. Effect of nanoplastics on the transport of platinum-based pharmaceuticals in water-saturated natural soil and their effect on a soil microbial community. Environmental Science: Nano. 7(10): 3178–3188. https://doi.org/10.1039/D0EN00651C.

Nielsen, F.H. 1996. How Should Dietary Guidance Be Given for Mineral Elements with Beneficial Actions or Suspected of Being Essential? The Journal of Nutrition. 126: 2377S-2385S. https://doi.org/10.1093/jn/126.suppl_9.2377S.

Nirel, P.M. and Revaclier, R. 1999. Assessment of sewage treatment plant effluents impact on river water quality using dissolved Rb/Sr ratio. Environmental Science and Technology. 33(12): 1996–2000. https://doi.org/10.1021/es981097g.

Nirel, P.M.V. and Lazzarotto, J. 2005. Testing of conductivity/calcium and rubidium/strontium ratios as indicators of the chemical stability of a river: comparison with a biological indicator. Water Science and Technology. 52(12): 291–296. https://doi.org/10.2166/wst.2005.0487.

Nur, T., Loganathan, P., Johir, M.A.H., Kandasamy, J. and Vigneswaran, S. 2018. Removing rubidium using potassium cobalt hexacyanoferrate in the membrane adsorption hybrid system. Separation and Purification Technology. 191: 286–294. https://doi.org/10.1016/j.seppur.2017.09.048.

Nuss, P. and Blengini, G.A. 2018. Towards better monitoring of technology critical elements in Europe: Coupling of natural and anthropogenic cycles. Science of The Total Environment. 613–614: 569–578. https://doi.org/10.1016/j.scitotenv.2017.09.117.

O'Driscoll, B. and González-Jiménez, J.M. 2016. Petrogenesis of the Platinum-Group Minerals. Reviews in Mineralogy and Geochemistry. 81(1): 489–578. https://doi.org/10.2138/rmg.2016.81.09

Olivarez, A.M. and Owen, R.M. 1989. REE/Fe variations in hydrothermal sediments: Implications for the REE content of seawater. Geochimica et Cosmochimica Acta. 53(3): 757–762. https://doi.org/10.1016/0016-7037(89)90019-7.

Oliveira, R.P., Benvenuti, J. and Espinosa, D.C.R. 2021. A review of the current progress in recycling technologies for gallium and rare earth elements from light-emitting diodes. Renewable and Sustainable Energy Reviews. 145: 111090. https://doi.org/10.1016/j.rser.2021.111090.

Oral, R., Bustamante, P., Warnau, M., D'Ambra, A., Guida, M. and Pagano, G. 2010. Cytogenetic and developmental toxicity of cerium and lanthanum to sea urchin embryos. Chemosphere. 81(2): 194–198. https://doi.org/10.1016/j.chemosphere.2010.06.057.

Orcutt, B.N., Bradley, J.A., Brazelton, W.J., Estes, E.R., Goordial, J.M., Huber, J.A., et al. 2020. Impacts of deep-sea mining on microbial ecosystem services. Limnology and Oceanography. 65(7): 1489–1510. https://doi.org/10.1002/lno.11403.

Osterauer, R., Haus, N., Sures, B. and Köhler, H.-R. 2009. Uptake of platinum by zebrafish (*Danio rerio*) and ramshorn snail (*Marisa cornuarietis*) and resulting effects on early embryogenesis. Chemosphere. 77(7): 975–982. https://doi.org/10.1016/j.chemosphere.2009.08.033.

Osterauer, R., Köhler, H.-R. and Triebskorn, R. 2010. Histopathological alterations and induction of hsp70 in ramshorn snail (*Marisa cornuarietis*) and zebrafish (*Danio rerio*) embryos after exposure to $PtCl_2$. Aquatic Toxicology. 99(1): 100–107. https://doi.org/10.1016/j.aquatox.2010.04.001.

Ouyang, Z., Huang, Q., Liu, B., Wu, H., Liu, T. and Liu, Y. 2019. Rubidium Chloride Targets Jnk/p38-Mediated NF-κB Activation to Attenuate Osteoclastogenesis and Facilitate Osteoblastogenesis. Frontiers in Pharmacology. 10: 584. https://doi.org/10.3389/fphar.2019.00584.

Pađan, J., Marcinek, S., Cindrić, A.-M., Layglon, N., Garnier, C., Salaün, P., et al. 2019. Determination of sub-picomolar levels of platinum in the pristine Krka River estuary (Croatia) using improved voltammetric methodology. Environmental Chemistry. 17(2): 77–84. https://doi.org/10.1071/EN19157.

Pagano, G., Guida, M., Tommasi, F. and Oral, R. 2015. Health effects and toxicity mechanisms of rare earth elements—Knowledge gaps and research prospects. Ecotoxicology and Environmental Safety. 115: 40–48. https://doi.org/10.1016/j.ecoenv.2015.01.030.

Pawlak, J., Łodyga-Chruścińska, E. and Chrustowicz, J. 2014. Fate of platinum metals in the environment. Journal of Trace Elements in Medicine and Biology. 28(3): 247–254. https://doi.org/10.1016/j.jtemb.2014.03.005.

Pekov, I.V. and Kononkova, N.N. 2010. Rubidium mineralization in rare-element granitic pegmatites of the Voron'i tundras, Kola Peninsula, Russia. Geochemistry International. 48(7): 695–713. https://doi.org/10.1134/S0016702910070062.

Perinajová, M., Janiga, M., Janiga, M. and Ballová, Z.K. 2020. Occurrence and vertical distribution of Ca, Cl, Cr, Fe, Mn, Mo, K, Rb, Sr, S, Sn, and Zn in the skull bones of alpine bullhead (*Cottus poecilopus*) in the West Carpathians. Environmental Science and Pollution Research. 27(29): 37114–37120. https://doi.org/10.1007/s11356-020-09752-7.

Peters, E.L., Schultz, I.R. and Newman, M.C. 1999. Rubidium and Cesium Kinetics and Tissue Distributions in Channel Catfish (*Ictalurus punctatus*). Ecotoxicology. 8(4): 287–300. https://doi.org/10.1023/A:1008981216690.

Petersen, S., Krätschell, A., Augustin, N., Jamieson, J., Hein, J.R. and Hannington, M.D. 2016. News from the seabed – Geological characteristics and resource potential of deep-sea mineral resources. Marine Policy. 70: 175–187. https://doi.org/10.1016/j.marpol.2016.03.012.

Pinto, J., Costa, M., Leite, C., Borges, C., Coppola, F., Henriques, B., et al. 2019. Ecotoxicological effects of lanthanum in *Mytilus galloprovincialis*: Biochemical and histopathological impacts. Aquatic Toxicology. 211: 181–192. https://doi.org/10.1016/j.aquatox.2019.03.017.

Pisoschi, A.M. and Pop, A. 2015. The role of antioxidants in the chemistry of oxidative stress: A review. European Journal of Medicinal Chemistry. 97: 55–74. https://doi.org/10.1016/j.ejmech.2015.04.040.

Pontoni, L., La Vecchia, C., Boguta, P., Sirakov, M., D'Aniello, E., Fabbricino, M., et al. 2022. Natural organic matter controls metal speciation and toxicity for marine organisms: a review. Environmental Chemistry Letters. 20(1): 797–812. https://doi.org/10.1007/s10311-021-01310-y.

Pöttgen, R., Jüstel, T. and Strassert, C.A. 2020. Rare Earth Chemistry. Walter de Gruyter GmbH & Co KG, Berlin.

Prodius, D., Gandha, K., Mudring, A.-V. and Nlebedim, I.C. 2019. Sustainable Urban Mining of Critical Elements from Magnet and Electronic Wastes. ACS Sustainable Chemistry and Engineering. https://doi.org/10.1021/acssuschemeng.9b05741.

PubChem. 2024. Rubidium. PubChem. Available at: https://pubchem.ncbi.nlm.nih.gov/element/Rubidium. Accessed on: 28 November 2024.

Purser, A., Marcon, Y., Hoving, H.-J.T., Vecchione, M., Piatkowski, U., Eason, D., et al. 2016. Association of deep-sea incirrate octopods with manganese crusts and nodule fields in the Pacific Ocean. Current Biology. 26(24): R1268–R1269. https://doi.org/10.1016/j.cub.2016.10.052.

Pusch, M., Bertorello, L. and Conti, F. 2000. Gating and Flickery Block Differentially Affected by Rubidium in Homomeric KCNQ1 and Heteromeric KCNQ1/KCNE1 Potassium Channels. Biophysical Journal. 78(1): 211–226. https://doi.org/10.1016/S0006-3495(00)76586-6.

Ravindra, K., Bencs, L. and Van Grieken, R. 2004. Platinum group elements in the environment and their health risk. Science of The Total Environment. 318(1): 1–43. https://doi.org/10.1016/S0048-9697(03)00372-3.

Reed, D.W., Fujita, Y., Daubaras, D.L., Jiao, Y. and Thompson, V.S. 2016. Bioleaching of rare earth elements from waste phosphors and cracking catalysts. Hydrometallurgy. 166: 34–40. https://doi.org/10.1016/j.hydromet.2016.08.006.

Rétif, J., Briant, N., Zalouk-Vergnoux, A., Le Monier, P., Sireau, T. and Poirier, L. 2024a. Distribution of rare earth elements and assessment of anthropogenic gadolinium in estuarine habitats: The case of Loire and Seine estuaries in France. Science of The Total Environment. 922: 171385. https://doi.org/10.1016/j.scitotenv.2024.171385.

Rétif, J., Zalouk-Vergnoux, A., Briant, N., François, Y. and Poirier, L. 2024b. Trophic dilution of rare earth elements along the food chain of the Seine estuary (France). Marine Pollution Bulletin. 206: 116671. https://doi.org/10.1016/j.marpolbul.2024.116671.

Revel, M., Drimmelen, C.K.E. van, Weltje, L., Hursthouse, A. and Heise, S. 2025. Effects of rare earth elements in the aquatic environment: Implications for ecotoxicological testing. Critical Reviews in Environmental Science and Technology.

Roberts, B.R., Doecke, J.D., Rembach, A., Yévenes, L.F., Fowler, C.J., McLean, C.A., et al. 2016. Rubidium and potassium levels are altered in Alzheimer's disease brain and blood but not in cerebrospinal fluid. Acta Neuropathologica Communications. 4(1): 119. https://doi.org/10.1186/s40478-016-0390-8.

Rochman, C.M., Manzano, C., Hentschel, B.T., Simonich, S.L.M. and Hoh, E. 2013. Polystyrene Plastic: A Source and Sink for Polycyclic Aromatic Hydrocarbons in the Marine Environment. Environmental Science and Technology. 47(24): 13976–13984. https://doi.org/10.1021/es403605f.

Rocio, M.A.R., Silva, M.M., Carvalho, P.S.L. de and Cardoso, J.G.R. 2012. Terras-raras: situação atual e perspectivas. BNDES Setorial. 35: 369–420.

Rodrigues, J.P., Santos-Echeandía, J., Chaves-Pozo, E., Campillo, J.A., Rocha-Santos, T., Duarte, A.C., et al. 2023. Interactive effects of palladium (Pd) and microplastics (MPs) on metal bioaccumulation and biological responses in the Mediterranean mussel, *Mytilus galloprovincialis*. Marine Pollution Bulletin. 194: 115284. https://doi.org/10.1016/j.marpolbul.2023.115284.

Romero-Freire, A., Santos-Echeandía, J., Neira, P. and Cobelo-García, A. 2019a. Less-Studied Technology-Critical Elements (Nb, Ta, Ga, In, Ge, Te) in the Marine Environment: Review on Their Concentrations in Water and Organisms. Frontiers in Marine Science. 6 https://doi.org/10.3389/fmars.2019.00532.

Romero-Freire, A., Joonas, E., Muna, M., Cossu-Leguille, C., Vignati, D.A.L. and Giamberini, L. 2019b. Assessment of the toxic effects of mixtures of three lanthanides (Ce, Gd, Lu) to aquatic biota. Science of The Total Environment. 661: 276–284. https://doi.org/10.1016/j.scitotenv.2019.01.155.

RSC, Royal Society of Chemistry. 2024a. Lutetium. Available at: https://periodic-table.rsc.org/element/71/lutetium. Accessed on: 28 November 2024.

RSC, Royal Society of Chemistry. 2024b. Lanthanum. Available at: https://periodic-table.rsc.org/element/57/lanthanum. Accessed on: 28 November 2024.

RSC, Royal Society of Chemistry. 2024c. Actinium. Available at: https://periodic-table.rsc.org/element/89/actinium#:~:text=The%20name%20is%20derived%20from,actinos'%2C%20meaning%20a%20ray. Accessed on: 28 November 2024.

RSC, Royal Society of Chemistry. 2025. Periodic Table. Available at: https://www.rsc.org/periodic-table. Accessed on: 15 September 2025.

Ruchter, N. and Sures, B. 2015. Distribution of platinum and other traffic related metals in sediments and clams (*Corbicula* sp.). Water Research. 70: 313–324. https://doi.org/10.1016/j.watres.2014.12.011.

Rudnick, R.L., Holland, H.D. and Rudnick, R.L. 2005. The Crust. Elsevier.

Rue, G.P. and McKnight, D.M. 2021. Enhanced Rare Earth Element Mobilization in a Mountain Watershed of the Colorado Mineral Belt with Concomitant Detection in Aquatic Biota: Increasing Climate Change-Driven Degradation to Water Quality. Environmental Science and Technology. https://doi.org/10.1021/acs.est.1c02958.

Sajkowska, S. and Leśniewska, B. 2024. Sorption of Platinum and Palladium on Polyethylene Microplastics in Natural Water. Molecules. 29(24): 5987. https://doi.org/10.3390/molecules29245987.

Saldaña-Serrano, M., Bastolla, C.L.V., Mattos, J.J., Lima, D., Freire, T.B., Nogueira, D.J., et al. 2022. Microplastics and linear alkylbenzene levels in oysters *Crassostrea gigas* driven by sewage contamination at an important aquaculture area of Brazil. Chemosphere. 307: 136039. https://doi.org/10.1016/j.chemosphere.2022.136039.

Sanders, M.J. and Kirschner, L.B. 1983. Potassium Metabolism in Seawater Teleosts: I. The Use of 86Rb as a Tracer for Potassium. Journal of Experimental Biology. 104(1): 15–28. https://doi.org/10.1242/jeb.104.1.15

Santos, A.C.S.S., Souza, L.A., Araujo, T.G., Rezende, C.E. de and Hatje, V. 2023. Fate and Trophic Transfer of Rare Earth Elements in a Tropical Estuarine Food Web. Environmental Science and Technology. https://doi.org/10.1021/acs.est.2c07726.

Sastri, V.R., Perumareddi, J.R., Rao, V.R., Rayudu, G.V.S. and Bünzli, J.-C.G. 2003. Modern Aspects of Rare Earths and their Complexes. Elsevier.

Scott, B.D.A. and Bray, W. 1980. Ancient Platinum Technology in South America: Its use by the Indians in Pre-Hispanic Times. Platinum Metals Review. 24(4): 147–157. https://doi.org/10.1595/003214080X244147157.

Seddon, K.R. 1996. Ruthenium: A dance to the music of time. Platinum Metals Review. 40(3): 128–134. https://doi.org/10.1595/003214096X403128134.

Seixas, S. and Pierce, G. 2005. Vanadium, rubidium and potassium in *Octopus vulgaris* (Mollusca: Cephalopoda). Scientia Marina. 69(2): 215–222. https://doi.org/10.3989/scimar.2005.69n2215.

Seymour, R.J. and O'Farrelly, J. 2012. Platinum-Group Metals. Kirk-Othmer Encyclopedia of Chemical Technology. https://doi.org/10.1002/0471238961.1612012019052513.a01

Shams, L., Turner, A., Millward, G.E. and Brown, M.T. 2014. Extra- and intracellular accumulation of platinum group elements by the marine microalga, *Chlorella stigmatophora*. Water Research. 50: 432–440. https://doi.org/10.1016/j.watres.2013.10.055.

Shan, Z.-Q., Shu, X.-Q., Feng, J.-F. and Zhou, W.-N. 2013. Modified calcination conditions of rare alkali metal Rb-containing muscovite ($KAl_2[AlSi_3O_{10}](OH)_2$). Rare Metals. 32(6): 632–635. https://doi.org/10.1007/s12598-013-0068-3.

Sharma, R. and Smith, S. 2019. Deep-Sea Mining and the Environment: An Introduction. Environmental Issues of Deep-Sea Mining: Impacts, Consequences and Policy Perspectives. 3–22. https://doi.org/10.1007/978-3-030-12696-4_1.

Shelby, J.E. 2021. Properties of potassium fluorogermanate glasses. Physics and Chemistry of Glasses—European Journal of Glass Science and Technology Part B. 62(1): 1–7. https://doi.org/10.13036/17533562.62.1.005.

Sholkovitz, E.R., Landing, W.M. and Lewis, B.L. 1994. Ocean particle chemistry: The fractionation of rare earth elements between suspended particles and seawater. Geochimica et Cosmochimica Acta. 58(6): 1567–1579. https://doi.org/10.1016/0016-7037(94)90559-2.

Singer, C., Zimmermann, S. and Sures, B. 2005. Induction of heat shock proteins (hsp70) in the zebra mussel *(Dreissena polymorpha)* following exposure to platinum group metals (platinum, palladium and rhodium): Comparison with lead and cadmium exposures. Aquatic Toxicology. 75(1): 65–75. https://doi.org/10.1016/j.aquatox.2005.07.004.

Smith, J.P., Boyd, T.J., Cragan, J. and Ward, M.C. 2021. Dissolved rubidium to strontium ratio as a conservative tracer for wastewater effluent-sourced contaminant inputs near a major urban wastewater treatment plant. Water Research. 205: 117691. https://doi.org/10.1016/j.watres.2021.117691.

Smrzka, D., Zwicker, J., Bach, W., Feng, D., Himmler, T., Chen, D., et al. 2019. The behavior of trace elements in seawater, sedimentary pore water, and their incorporation into carbonate minerals: a review. Facies. 65(4): 41. https://doi.org/10.1007/s10347-019-0581-4.

Sousa Filho, P.C. de and Serra, O.A. 2014. Terras raras no Brasil: histórico, produção e perspectivas. Química Nova. 37: 753–760. https://doi.org/10.5935/0100-4042.20140121.

Souza, L.A., Santos, A.C.S.S., Leão, J.M., Schaeppi, O.C. and Hatje, V. 2022. Occurrence and contents of trace metals and rare earth elements on plastic pellets. Marine Pollution Bulletin. 185: 114261. https://doi.org/10.1016/j.marpolbul.2022.114261.

Soyol-Erdene, T.O. and Huh, Y. 2012. Dissolved platinum in major rivers of East Asia: Implications for the oceanic budget. Geochemistry, Geophysics, Geosystems. 13(6). https://doi.org/10.1029/2012GC004102.

Squadrone, S., Brizio, P., Stella, C., Mantia, M., Favaro, L., Biancani, B., et al. 2020. Differential bioaccumulation of trace elements and rare earth elements in the muscle, kidneys, and liver of the invasive indo-pacific lionfish (*Pterois* spp.) from Cuba. Biological Trace Element Research. 196(1): 262–271. https://doi.org/10.1007/s12011-019-01918-w.

Su, Y., Chen, L.-J., He, J.-R., Yuan, X.-J., Cen, Y.-L., Su, F.-X., et al. 2011. Urinary rubidium in breast cancers. Clinica Chimica Acta. 412(23): 2305–2309. https://doi.org/10.1016/j.cca.2011.08.035.

Sysolyatina, M.A. and Olkova, A.S. 2022. Sources of rare earth elements in the environment and their impact on living organisms. Environmental Reviews. https://doi.org/10.1139/er-2022-0081.

Tai, P., Zhao, Q., Su, D., Li, P. and Stagnitti, F. 2010. Biological toxicity of lanthanide elements on algae. Chemosphere. 80(9): 1031–1035. https://doi.org/10.1016/j.chemosphere.2010.05.030.

Takashima, R., Selby, D., Yamanaka, T., Kuwahara, Y., Nakamura, H., Sawada, K., et al. 2024. Large igneous province activity drives oceanic anoxic event 2 environmental change across eastern Asia. Communications Earth & Environment. 5(1): 85. https://doi.org/10.1038/s43247-024-01214-z.

Talan, D. and Huang, Q. 2022. A review of environmental aspect of rare earth element extraction processes and solution purification techniques. Minerals Engineering. 179: 107430. https://doi.org/10.1016/j.mineng.2022.107430.

Tang, H., Zhao, L., Sun, W., Hu, Y., Ji, X., Han, H., et al. 2018. Extraction of rubidium from respirable sintering dust. Hydrometallurgy. 175: 144–149. https://doi.org/10.1016/j.hydromet.2017.11.003.

Tao, Y., Shen, L., Feng, C., Yang, R., Qu, J., Ju, H., et al. 2022. Distribution of rare earth elements (REEs) and their roles in plant growth: A review. Environmental Pollution. 298: 118540. https://doi.org/10.1016/j.envpol.2021.118540.

Tavakoli Mohammadi, M.R., Javad Koleini, S.M., Javanshir, S., Abolghasemi, H. and Abdollahy, M. 2015. Solvent extraction of rubidium from gold waste using conventional SX and new CFE methods. Rare Metals. 34(11): 818–828. https://doi.org/10.1007/s12598-015-0463-z.

Thomas, P.J., Carpenter, D., Boutin, C. and Allison, J.E. 2014. Rare earth elements (REEs): Effects on germination and growth of selected crop and native plant species. Chemosphere. 96: 57–66. https://doi.org/10.1016/j.chemosphere.2013.07.020.

Thurber, A.R., Sweetman, A.K., Narayanaswamy, B.E., Jones, D.O.B., Ingels, J. and Hansman, R.L. 2014. Ecosystem function and services provided by the deep sea. Biogeosciences. 11(14): 3941–3963. https://doi.org/10.5194/bg-11-3941-2014.

Tian, N., Wu, J., Wang, J. and Dai, W. 2019. Development of a Novel Core–Shell Magnetic Fe_3O_4@CMC@ZIF-8-OH Composite with Outstanding Rubidium-Ion Capacity. Journal of Chemical & Engineering Data. 64(12): 5716–5724. https://doi.org/10.1021/acs.jced.9b00708.

Tian, L., Gong, A., Wu, X., Li, J., Liu, C., Yu, X., et al. 2020. Non-isothermal kinetic studies of rubidium extraction from muscovite using a chlorination roasting-water leaching process. Powder Technology. 373: 362–368. https://doi.org/10.1016/j.powtec.2020.06.015.

Toyoda, K., Nakamura, Y. and Masuda, A. 1990. Rare earth elements of Pacific pelagic sediments. Geochimica et Cosmochimica Acta. 54(4): 1093–1103. https://doi.org/10.1016/0016-7037(90)90441-M.

Trifuoggi, M., Pagano, G., Guida, M., Palumbo, A., Siciliano, A., Gravina, M., et al. 2017. Comparative toxicity of seven rare earth elements in sea urchin early life stages. Environmental Science and Pollution Research. 24(25): 20803–20810. https://doi.org/10.1007/s11356-017-9658-1.

Trimmel, S., Spörl, P., Haluza, D., Lashin, N., Meisel, T.C., Pitha, U., et al. 2024. Green and blue infrastructure as model system for emissions of technology-critical elements. Science of The Total Environment. 934: 173364. https://doi.org/10.1016/j.scitotenv.2024.173364.

Trombini, C., Garcia da Fonseca, T., Morais, M., Rocha, T.L., Blasco, J. and Bebianno, M.J. 2016. Toxic effects of cisplatin cytostatic drug in mussel *Mytilus galloprovincialis*. Marine Environmental Research. 119: 12–21. https://doi.org/10.1016/j.marenvres.2016.05.004.

Tunnicliffe, V., Metaxas, A., Le, J., Ramirez-Llodra, E. and Levin, L.A. 2020. Strategic Environmental Goals and Objectives: Setting the basis for environmental regulation of deep seabed mining. Marine Policy. 114: 103347. https://doi.org/10.1016/j.marpol.2018.11.010.

Turner, A., Lewis, M.S., Shams, L. and Brown, M.T. 2007. Uptake of platinum group elements by the marine macroalga, *Ulva lactuca*. Marine Chemistry. 105(3): 271–280. https://doi.org/10.1016/j.marchem.2007.02.009.

USGS, U.S. Geological Survey. 2024. Mineral Commodity Summaries 2024. U.S. Geological Survey, Reston, Virginia, 216. https://doi.org/10.5066/P144BA54.

Usuda, K., Kono, R., Ueno, T., Ito, Y., Dote, T., Yokoyama, H., et al. 2014. Risk Assessment Visualization of Rubidium Compounds: Comparison of Renal and Hepatic Toxicities, *In vivo*. Biological Trace Element Research. 159(1): 263–268. https://doi.org/10.1007/s12011-014-9937-3.

Van Dover, C.L. 2011. Mining seafloor massive sulphides and biodiversity: what is at risk? ICES Journal of Marine Science. 68(2): 341–348. https://doi.org/10.1093/icesjms/fsq086

Voncken, J.H.L. 2016. The Rare Earth Elements. Springer. https://doi.org/10.1007/978-3-319-26809-5.

Vysetti, B. 2023. Deep-sea mineral deposits as a future source of critical metals, and environmental issues—A brief review. Minerals and Mineral Materials. 2(2): N/A-N/A. https://doi.org/10.20517/mmm.2022.12.

Wang, J., Che, D. and Qin, W. 2015. Extraction of rubidium by *t*-BAMBP in cyclohexane. Chinese Journal of Chemical Engineering. 23(7): 1110–1113. https://doi.org/10.1016/j.cjche.2015.04.005.

Wang, Z., Yin, L., Xiang, H., Qin, X. and Wang, S. 2019. Accumulation patterns and species-specific characteristics of yttrium and rare earth elements (YREEs) in biological matrices from Maluan Bay, China: Implications for biomonitoring. Environmental Research. 179: 108804. https://doi.org/10.1016/j.envres.2019.108804.

Wang, X.-N., Gu, Y.-G. and Wang, Z.-H. 2022. Rare earth elements in different trophic level marine wild fish species. Environmental Pollution. 292: 118346. https://doi.org/10.1016/j.envpol.2021.118346.

Wen, Y., Hu, L., Boxleiter, A., Li, D. and Tang, Y. 2023. Rare Earth Elements Recovery and Waste Management of Municipal Solid Waste Incineration Ash. ACS Sustainable Resource Management. https://doi.org/10.1021/acssusresmgt.3c00026.

Weng, Z., Jowitt, S.M., Mudd, G.M. and Haque, N. 2015. A detailed assessment of global rare earth element resources: Opportunities and challenges. Economic Geology. 110(8): 1925–1952. https://doi.org/10.2113/econgeo.110.8.1925.

Wilburn, D.R. 2012. Global exploration and production capacity for platinum-group metals from 1995 through 2015. Scientific Investigations Report. https://doi.org/10.3133/sir20125164.

Williams, L.R. and Leggett, R.W. 1987. The distribution of intracellular alkali metals in Reference Man. Physics in Medicine and Biology. 32(2): 173. https://doi.org/10.1088/0031-9155/32/2/002.

Wollaston, W.H. 1809. X. On platina and native palladium from Brasil. Philosophical Transactions of the Royal Society of London. 99: 189–194. https://doi.org/10.1098/rstl.1809.0012.

Wosnick, N., Chaves, A.P., Leite, R.D., Nunes, J.L.S., Saint'Pierre, T.D., Willmer, I.Q., et al. 2021. Nurse sharks, space rockets and cargo ships: Metals and oxidative stress in a benthic, resident and large-sized mesopredator, *Ginglymostoma cirratum*. Environmental Pollution. 288: 117784. https://doi.org/10.1016/j.envpol.2021.117784.

Wosnick, N., Chaves, A.P., Giareta, E.P., Leite, R.D., Saint'Pierre, T.D. and Hauser-Davis, R.A. 2024. From screens to seas: Tech contaminants in tiger sharks. Environmental Pollution. 360: 124690. https://doi.org/10.1016/j.envpol.2024.124690.

Wright, D.A. 1995. Trace metal and major ion interactions in aquatic animals. Marine Pollution Bulletin. 31(1): 8–18. https://doi.org/10.1016/0025-326X(95)00036-M.

Wu, Q., Qi, J. and Xia, X. 2017. Long-term variations in sediment heavy metals of a reservoir with changing trophic states: Implications for the impact of climate change. Science of The Total Environment. 609: 242–250. https://doi.org/10.1016/j.scitotenv.2017.04.041.

Wu, L., Li, G., Zhu, X., Jia, G. and Yan, W. 2024. Asian winter monsoon controls marine primary productivity in north Arabian Sea during the Holocene. Palaeogeography, Palaeoclimatology, Palaeoecology. 648: 112287. https://doi.org/10.1016/j.palaeo.2024.112287.

Xia, J. and Ghahreman, A. 2023. Platinum group metals recycling from spent automotive catalysts: metallurgical extraction and recovery technologies. Separation and Purification Technology. 311: 123357. https://doi.org/10.1016/j.seppur.2023.123357.

Xing, P., Wang, C., Wang, L., Ma, B., Chen, Y. and Wang, G. 2018. Clean and efficient process for the extraction of rubidium from granitic rubidium ore. Journal of Cleaner Production. 196: 64–73. https://doi.org/10.1016/j.jclepro.2018.06.041.

Xing, P., Wang, C., Chen, Y. and Ma, B. 2021. Rubidium extraction from mineral and brine resources: A review. Hydrometallurgy. 203: 105644. https://doi.org/10.1016/j.hydromet.2021.105644.

Xu, T., and Peng, H. 2009. Formation cause, composition analysis and comprehensive utilization of rare earth solid wastes. Journal of Rare Earths. 27(6): 1096–1102. doi:10.1016/S1002-0721(08)60394-4.

Yamaguchi, S., Miura, C., Ito, A., Agusa, T., Iwata, H., Tanabe, S., et al. 2007. Effects of lead, molybdenum, rubidium, arsenic and organochlorines on spermatogenesis in fish: Monitoring at Mekong Delta area and *in vitro* experiment. Aquatic Toxicology. 83(1): 43–51. https://doi.org/10.1016/j.aquatox.2007.03.010.

Yang, L., Wang, X., Nie, H., Shao, L., Wang, G. and Liu, Y. 2016. Residual levels of rare earth elements in freshwater and marine fish and their health risk assessment from Shandong, China. Marine Pollution Bulletin. 107(1): 393–397. https://doi.org/10.1016/j.marpolbul.2016.03.034.

Ye, X., Wu, Z., Li, W., Liu, H., Li, Q., Qing, B., et al. 2009. Rubidium and cesium ion adsorption by an ammonium molybdophosphate–calcium alginate composite adsorbent. Colloids and Surfaces A: Physicochemical and Engineering Aspects. 342(1): 76–83. https://doi.org/10.1016/j.colsurfa.2009.04.011.

Yokoi, K., Kimura, M. and Itokawa, Y. 1996. Effect of low dietary rubidium on plasma biochemical parameters and mineral levels in rats. Biological Trace Element Research. 51(2): 199–208. https://doi.org/10.1007/BF02785438.

Yokoi, K. 1997. A study on possible essentiality nature of tin and rubidium. Nippon Eiyo Shokuryo Gakkaishi. 50(1): 15–20. https://doi.org/10.4327/jsnfs.50.15.

Zeng, Q., Huang, L., Ouyang, D., Hu, Y., Zhong, H. and He, Z. 2019. Process optimization on the extraction of rubidium from rubidium-bearing biotite. Minerals Engineering. 137: 87–93. https://doi.org/10.1016/j.mineng.2019.03.020.

Zepf, V. 2013. Rare earth elements: What and where they are. *In*: Rare Earth Elements: A New Approach to the Nexus of Supply, Demand and Use: Exemplified along the Use of Neodymium in Permanent Magnets. 11–39. Springer Theses. Springer, Berlin, Heidelberg. https://doi.org/10.1007/978-3-642-35458-8_2.

Zhang, K., Zocher, A.-L. and Bau, M. 2025. Rare earth element and yttrium behaviour during metabolic transfer and biomineralisation in the marine bivalve *Mytilus edulis*: Evidence for a (partially) biological origin of REY anomalies in mussel shells. Science of The Total Environment. 958: 178056. https://doi.org/10.1016/j.scitotenv.2024.178056.

Zhao, C.-M., Campbell, P.G.C. and Wilkinson, K.J. 2016. When are metal complexes bioavailable? Environmental Chemistry. 13(3): 425–433. https://doi.org/10.1071/EN15205.

Zheng, S., Li, P., Tian, L., Cao, Z., Zhang, T., Chen, Y., et al. 2016. A chlorination roasting process to extract rubidium from distinctive kaolin ore with alternative chlorinating reagent. International Journal of Mineral Processing. 157: 21–27. https://doi.org/10.1016/j.minpro.2016.09.006.

Zhou, L., Yuan, T., Li, R., Zhong, Y. and Lei, X. 2015. Extraction of rubidium from kaolin clay waste: Process study. Hydrometallurgy. 158: 61–67. https://doi.org/10.1016/j.hydromet.2015.10.010.

Zimmermann, S., Alt, F., Messerschmidt, J., von Bohlen, A., Taraschewski, H. and Sures, B. 2002. Biological availability of traffic-related platinum-group elements (palladium, platinum, and rhodium) and other metals to the zebra mussel (*Dreissena polymorpha*) in water containing road dust. Environmental Toxicology and Chemistry. 21(12): 2713–2718. https://doi.org/10.1002/etc.5620211226

Zimmermann, S., Baumann, U., Taraschewski, H. and Sures, B. 2004. Accumulation and distribution of platinum and rhodium in the European eel *Anguilla anguilla* following aqueous exposure to metal salts. Environmental Pollution. 127(2): 195–202. https://doi.org/10.1016/j.envpol.2003.08.006.

Zimmermann, S., Wolff, C. and Sures, B. 2017. Toxicity of platinum, palladium and rhodium to *Daphnia magna* in single and binary metal exposure experiments. Environmental Pollution. 224: 368–376. https://doi.org/10.1016/j.envpol.2017.02.016.